SHIP (S-BAND)

SIGNALS FROM EARTH (VHF RELAYED)

VOICE/DATA (VHF)

VOICE (VHF)

SIGNALS FROM EARTH (VHF RELAYED)

VOICE/DATA FROM BOTH ASTRONAUTS (VHF)

COMMUNICATIONS
UNIT (LCRU)

The Observer's Pocket Series

MANNED SPACEFLIGHT

The Observer Books

A POCKET REFERENCE SERIES

COVERING NATURAL HISTORY, TRANSPORT,

THE ARTS ETC

The Observer's Book of
MANNED SPACEFLIGHT

REGINALD TURNILL

WITH 14 COLOUR AND OVER 70
BLACK AND WHITE ILLUSTRATIONS

FREDERICK WARNE & CO LTD
FREDERICK WARNE & CO INC
LONDON: NEW YORK

LIBRARY OF CONGRESS CATALOG
CARD NO. 72–81147

ISBN 0 7232 1510 3

Printed in Great Britain by
William Clowes & Sons, Limited,
London, Beccles and Colchester
1045.572

CONTENTS

LIST OF COLOUR PLATES

ACKNOWLEDGMENTS

Much research for this book had already been done for the same author's *Language of Space: a Dictionary of Astronautics* (Cassells, London; John Day, New York). The author is grateful for the ready assistance provided by the US National Aeronautics and Space Administration (NASA), and for the information available in the News Reference books provided jointly by them and the space contractors for the use of correspondents. Other invaluable sources include the Press Kits issued for individual missions by NASA and the prime contractor companies (North American Rockwell; Grumman Aircraft; McDonnell Douglas; the Boeing Company; the Chrysler Corporation; TRW Inc; and many more). Most of the US photographs are provided by NASA— many having been taken by the astronauts themselves; others have been issued by the contractors and the US Embassy. Pictures of Vostok, Soyuz, etc. are provided by Tass, Novosti and Graham Turnill. The thanks of the publishers and author are extended to all of them; additionally, the author's gratitude is also due to Margaret Turnill for her painstaking research, and her typing and checking of the manuscript; and to Anne Emerson of Frederick Warne for the final editing which resulted in the elimination of many mistakes.

INTRODUCTION

The aim of this first edition of *The Observer's Book of Manned Spaceflight* is to provide an interesting and entertaining introduction, as well as a comprehensive reference book in handy form for the established student of the subject.

The Space Logs with which it begins enable the reader, if he wishes, to study the subject chronologically by turning at once to Vostok 1, and the historic first spaceflight by Yuri Gagarin in 1961; equally the steady American progress towards the first moonlanding can be traced through the Mercury, Gemini and Apollo projects. To enable the book to be used for quick reference, however, the projects appear in alphabetical order.

Although man's first experimental satellite, the Russian Sputnik 1, went into orbit as recently as October 4, 1957, it has not been possible to cover the whole subject in one small volume; it is hoped this will be achieved in due course with the production of a complementary volume entitled *Unmanned Spaceflight*. This has meant, in the case of America, setting aside the unmanned projects such as Ranger, Lunar Orbiter and Surveyor which paved the way for the first men to land on the moon. It has meant, too, the omission of America's Mariner, Pioneer and Viking programmes; and of Russia's Luna, Venera and Mars probes, which are already exploring Mars in detail, and making the first giant steps towards similar remotely controlled studies of the outer planets, Jupiter and Saturn.

Past experience suggests that these programmes

will lead on to the first manned planetary expeditions—at first without any attempt at landing—much sooner than is at present anticipated. So, although America's series of Apollo moonlandings is coming to an end as this first edition of *Manned Spaceflight* goes to press, there is plenty of excitement ahead both for students of space and future editions of this book. Evidence of that is to be found in the picture of Skylab on the dust cover, and the full details in the text of America's first 'orbiting workshop'—really a highly sophisticated space station—which should be operating throughout 1973. The latest details are also included of the even more exciting, (though constantly changing) plans for a Space Shuttle, which will finally combine the performance of aircraft and spacecraft in one vehicle.

It would be pleasant to be able to include similar signposts to the space adventures undoubtedly being planned by Russia; they still prefer, however, to maintain strict secrecy about their future plans for both manned and unmanned flight. It is a consolation that a careful study of the past achievements of the rival Soviet and American programmes, together with detailed knowledge of America's future plans, make it possible to build up forecasts of what Russia is likely to do.

For instance, it seems certain that once the American moonlandings have ceased, at least for the time being, the Russians will feel the time has come to start their own; and undoubtedly they will want their expeditions to be on a much more ambitious scale, with manned visits to the lunar poles and the moon's farside which were far beyond the reach of the Apollo men.

While America's Skylab cluster, operated by rotating crews of astronauts, studies possible ways of harnessing the sun's power and of controlling the steady pollution of earth, we shall almost certainly see Russia, with new versions of her Salyut space stations, moving towards similar experiments. And experience gained in assembling these space stations in orbit will bring nearer the day when man sets out for the planets, since it is only by such assembly procedures that huge vehicles—containing about 11 tons (11,180 kg) of hydrogen, food and water for each of three men for every year of flight—can be launched on interplanetary expeditions.

The first orbital flight of the Space Shuttle, which America is now developing to succeed Apollo, is not likely to take place until 1978. But the gap between the last US moonlanding at the end of 1972 and the start of that new era of manned space travel 6 years later will be satisfyingly filled, for, besides the Skylab missions of 1973, several ambitious flights are being planned with left-over Apollo equipment. And, in addition to the always unannounced Soviet 'spectaculars', it is almost certain that around June 1975 an Apollo spacecraft will link up with a Salyut/Soyuz space station. For the first time astronauts and cosmonauts will shake hands in space and participate in a joint mission.

LIST OF ABBREVIATIONS

ALSEP	Apollo Surface Experimental Package
AM	Airlock Module
ATM	Apollo Telescope Mount
BST	British Summer Time
CM	Command Module
CMP	Command Module Pilot
CSM	Command and Service Module
ECS	Environmental Control System
EDT	Eastern Daylight Time
EVA	Extravehicular Activity
G or G Force	The force exerted upon an object by gravity. '1G' is the measure of gravitational pull required to accelerate a body at the rate of 32·16 feet per second
GMT	Greenwich Mean Time
GT	Gemini-Titan flights
GTA	Gemini-Titan-Agena flights
ICBM	Intercontinental Ballistic Missile
IU	Instrument Unit
LCC	Launch Control Centre (Cape Kennedy)
LM	Lunar Module
LMP	Lunar Module Pilot
LOI	Lunar Orbit Insertion
LOX	Liquid Oxygen
LRV	Lunar Roving Vehicle
MA	Mercury-Atlas flights
MC	Mission Control
MDA	Multiple Docking Adapter
MDS	Malfunction Detection Systems
MESA	Modularized Equipment Stowage Assembly
MET	Mission Event Timer

MOL	Manned Orbiting Laboratory
MR	Mercury-Redstone flights
MSC	Manned Spacecraft Centre (Houston)
MSS	Mobile Service Structure
NASA	National Aeronautics and Space Administration (US)
OWS	Orbiting Workshop
PLSS	Portable Life Support Systems
psi	pounds per square inch
RCS	Reaction Control System
RV	Rendezvous
SIM	Scientific Instrument Module
SIMBAY	Bay containing Scientific Instrument Module
SM	Service Module
SP	Service Propulsion
SPS	Service Propulsion System
STS	Space Transportation System
SWS	Saturn Workshop
T	Time, the moment of ignition counting as zero or $T-0$. Time is given in minus quantities before launch, and plus quantities after.
TACS	Thruster Attitude Control System
TEI	Transearth Insertion
TLI	Translunar Insertion
UDMH	Unsymmetrical Dimethylhydrazine
USAF	United States Air Force
USMC	United States Marine Corps
USN	United States Navy
VAB	Vehicle Assembly Building

13

SPACE LOGS
US Manned Space Flights

Spacecraft	Launch Date	Astronauts	Orbits	Flt Time Hrs Mins		Highlights
Mercury 3	May 5, 1961	Alan Shepard	Sub-orbital	0.	15	First American in space
Mercury 4	July 21, 1961	Virgil Grissom	Sub-orbital	0.	16	Capsule sank
Mercury 6	Feb. 20, 1962	John Glenn	3	4.	55	First American in orbit
Mercury 7	May 24, 1962	M. Scott Carpenter	3	4.	56	Landed 250 miles (402 km) from target
Mercury 8	Oct. 3, 1962	Walter Schirra	6	9.	13	Landed 5 miles (8 km) from target
Mercury 9	May 15, 1963	L. Gordon Cooper	22	34.	20	First long flight by an American
Gemini 3	Mar. 23, 1965	Virgil Grissom John Young	3	4.	52	First manned orbital manoeuvres
Gemini 4	June 3, 1965	James McDivitt Edward White	62	97.	56	21-min 'spacewalk' (White)
Gemini 5	Aug. 21, 1965	L. Gordon Cooper Charles Conrad	120	190.	55	First extended manned flight
Gemini 7	Dec. 4, 1965	Frank Borman James Lovell	206	330.	35	Longest US spaceflight

Mission	Date	Crew	Orbits			Notes
Gemini 6	Dec. 15, 1965	Walter Schirra Thomas Stafford	16	25.	51	RV to 6 ft (1·8 m) of Gemini 7
Gemini 8	Mar. 16, 1966	Neil Armstrong David Scott	6	10.	41	First docking; emergency splash-down
Gemini 9	June 3, 1966	Thomas Stafford Eugene Cernan	45	72.	21	2-hr spacewalk (Cernan)
Gemini 10	July 18, 1966	John Young Michael Collins	43	70.	47	RV with 2 targets; Agena package retrieved
Gemini 11	Sept. 12, 1966	Charles Conrad Richard Gordon	44	71.	17	RV and docking
Gemini 12	Nov. 11, 1966	James Lovell Edwin Aldrin	59	94.	34	Dockings; 3 spacewalks
Apollo 7	Oct. 11, 1968	Walter Schirra Donn Eisele Walter Cunningham	163	260.	9	First manned Apollo flight
Apollo 8	Dec. 21, 1968	Frank Borman James Lovell William Anders	Lunar Orbits 10	147.	0	First manned flight around moon
Apollo 9	Mar. 3, 1969	James McDivitt David Scott Russell Schweickart	151	241.	1	Docking with Lunar Module
Apollo 10	May 18, 1969	Thomas Stafford Eugene Cernan John Young	Lunar Orbits 31	192.	3	Descent to within 9 miles of moon

Mission	Date	Crew	CM L. Orbits			Notes
Apollo 11	July 16, 1969	Neil Armstrong Edwin Aldrin Michael Collins	CM L. Orbits 31	195.	18	Armstrong and Aldrin land on moon; 44 lb (20 kg) samples
Apollo 12	Nov. 14, 1969	Charles Conrad Richard Gordon Alan Bean	CM L. Orbits 45	244.	36	2 EVAs, total 7 hrs 39 mins; 75 lb (34 kg) samples
Apollo 13	Apr. 13, 1970	James Lovell John Swigert Fred Haise		142.	55	Mission aborted following oxygen tank explosion
Apollo 14	Jan. 31, 1971	Alan Shepard Stuart Roosa Edgar Mitchell	CM L. Orbits 34	216.	02	2 EVAs, total 9 hrs 25 mins; 98 lb (44 kg) samples
Apollo 15	July 26, 1971	David Scott James Irwin Alfred Worden	CM L. Orbits 74	295.	12	3 EVAs, total 18 hrs 36 mins; 173 lb (78 kg) samples
Apollo 16	Apr. 16, 1972	John Young Thos Mattingly Charles Duke	64	265.	51	3 EVAs, total 20 hrs 14 mins; 215 lb (97·5 kg) samples
Apollo 17	Dec. 6, 1972	Eugene Cernan Ronald Evans Harrison Schmitt				

SPACE LOGS
Soviet Manned Flights

Spacecraft	Launch Date	Cosmonauts	Orbits	Flt Time Hrs Mins		Highlights
Vostok 1	Apr. 12, 1961	Yuri Gagarin	1	1.	48	First man in space
Vostok 2	Aug. 6, 1961	Herman Titov	17	25.	18	First day in space
Vostok 3	Aug. 11, 1962	Andrian Nikolayev	64	94.	27	First double flight
Vostok 4	Aug. 12, 1962	Pavel Popovich	48	70.	29	
Vostok 5	June 14, 1963	Valeri Bykovsky	81	119.	06	Second group flight
Vostok 6	June 16, 1963	Valentina Tereshkova	48	70.	50	First woman in space
Voskhod 1	Oct. 12, 1964	Vladimir Komarov Konstantin Feoktistov Boris Yegorov	16	24.	17	First three-man craft
Voskhod 2	Mar. 18, 1965	Pavel Belyayev Alexei Leonov	17	26.	02	First spacewalk (10 min) by Leonov
Soyuz 1	Apr. 23, 1967	Vladimir Komarov	18	26.	45	Re-entry parachute snarled; Komarov killed
Soyuz 3	Oct. 26, 1968	Giorgi Beregovoi	64	94.	51	RV practice with unmanned Soyuz 2

Soyuz 4	Jan. 14, 1969	Vladimir Shatalov	48	71.	14	First docking of two manned craft; Khruonov and Yeliseyev returned in Soyuz 4
Soyuz 5	Jan. 15, 1969	Boris Volynov Yevgeny Khrunov Alexei Yeliseyev	50	72.	46	
Soyuz 6	Oct. 11, 1969	Georgi Shonin Valeri Kubasov	80	118.	42	
Soyuz 7	Oct. 12, 1969	Anatoli Filipchenko Vladislav Volkov Viktor Gorbatko	80	118.	41	30 manual control manoeuvres; first space welding
Soyuz 8	Oct. 13, 1969	Vladimir Shatalov Alexei Yeliseyev	80	118.	41	
Soyuz 9	June 1, 1970	Andrian Nikolayev Vitali Sevastyanov	285	424.	59	Endurance record (nearly 18 days)
Salyut	Apr. 19, 1971	Orbital Science Station				Destroyed Oct. 11, 1971 after approx. 2,800 orbits
Soyuz 10	Apr. 23, 1971	Vladimir Shatalov Alexei Yeliseyev Nikolai Rukavish- nikov	30	48.	45	Docked with Salyut but no entry made
Soyuz 11	June 6, 1971	Georgi Dobrovolsky Vladislav Volkov Viktor Patsayev	380	569.	40	23 days spent in Salyut; crew killed on re-entry

18

Comparison of US and USSR Manned Flights to end of May 1972 (up to and including Apollo 16 and Soyuz 11)

	US	USSR
Total flights	26	18
Spacemen Involved	32	25
Spacecraft Hours	3223 hrs 51 mins	2097 hrs 21 mins
Cumulative Man Hours	8593 hrs 54 mins	4397 hrs 41 mins
Earth Orbits	968	1407
Lunar Orbits	289	0

Note: Totals are based on the Space Logs above. It should be noted that official NASA figures often vary slightly, and that there is constant confusion between orbits and revolutions. (A revolution is 6 minutes longer than an orbit.) Soviet totals are based on official figures when available, and estimates when they were not. The US has been credited with two earth orbits for each moonflight.

Apollo 11 mounted on launch complex 39A

MANNED
SPACE VEHICLES

APOLLO—
US MANNED SPACECRAFT

Apollo 11 Launch Weights

CM	12,253 lb	(5558 kg)
SM	51,156 lb	(23,204 kg)
LM Adapter	4049 lb	(1837 kg)
LM	33,277 lb	(15,094 kg)
Escape System	8910 lb	(4041 kg)
	109,645 lb	(49,734 kg)

Spacecraft Weight into Earth Orbit:
    100,745 lb (45,697 kg)
CM Weight at Splashdown:
    10,971 lb (4976 kg)

History   A three-man spacecraft was first proposed by NASA in July 1960, for earth orbital
and circumlunar flights, to be launched by a
Saturn 1-type rocket with 1·5 million lb (680,400
kg) thrust.   On May 25, 1961 President John F.
Kennedy proposed that the US should establish as
a national goal a manned landing on the moon by
the end of the decade, and this in turn led to the
development of Saturn 5, with 7·5 million lb
(3·40 million kg) of thrust.   North American
Rockwell was selected as prime contractor for the
Apollo Command and Service Modules, and
Grumman Aircraft for the Lunar Module.   On
May 28, 1964 a Saturn 1 placed the first Apollo
Command Module in orbit from Cape Kennedy;

parallel development of the spacecraft and rockets was making remarkable progress until, on January 27, 1967, an electrical arc from wiring in a spacecraft being ground-tested at Cape Kennedy started a fire which became catastrophic in the 100% oxygen-atmosphere. Astronauts Grissom, White and Chaffee were burned to death, and the first manned flight was delayed for eighteen months. On November 9, 1967 the first unmanned test of the combined Apollo/Saturn 5 vehicles—designated Apollo 4—was successfully accomplished. Apollo 5, in January 1968, successfully tested the Lunar Module systems, including firings in earth orbit of both the ascent and descent propulsion systems; in April 1968 Apollo 6, the second unmanned test of the combined Apollo/Saturn 5, was only partially successful since vertical oscillations, or 'Pogo' effects, were encountered with the first stage. Nevertheless, the first manned flight, Apollo 7, went ahead in October 1968 (Plate 1), and the first lunar landing by Apollo 11 was astonishingly achieved only 9 months later. A mission-by-mission account follows on pages 35–69.

Spacecraft Description The complete Apollo spacecraft is 82 ft (25 m) tall, and consists of: 1) the Command Module; 2) the Service Module; 3) the Lunar Module; 4) the Launch Escape System; 5) the Spacecraft Lunar Module Adapter.

The Launch Escape System consists of a 33-ft (10-m) tower weighing 8000 lb (3629 kg) and a solid-rocket motor $15\frac{1}{2}$ ft (4·72 m) long, providing 147,000 lb (66,675 kg) of thrust. If a fire or abort situation occurs during the countdown or in the first 100 seconds of lift-off, the Commander presses the abort button and the Escape Tower lifts the spacecraft about 1 mile (1·6 km) clear of the pad and rocket; the main parachute system is then used for descent. *The Spacecraft Lunar Module Adapter* serves as a smooth aerodynamic enclosure, linking spacecraft and launch vehicle during lift-off, and protecting the Lunar Module, which is extracted from it during transposition and docking, shortly after the spacecraft leaves earth orbit on its journey to the moon.

Command Module The control centre and living quarters for the three-man crew; 11 ft 5 in. (3·48 m) from nose to heatshield. One man spends the entire mission in it; the others leave it only for the lunar landing; it is the only part finally to return to earth. Lift-off weight has now risen to about 12,800 lb (5806 kg) including crew; splashdown weight about 11,700 lb (5307 kg). As on an aircraft, the Commander, who normally operates the flight controls, is on the left; the CM pilot, responsible for guidance and navigation, is on the centre couch; the LM pilot, responsible for management of subsystems, is on the right. Their couches face the main display console, nearly 7 ft

Launch
Escape
System

Boost
Protective
Cover

Command
Module

Service
Module

Adapter

Lunar
Module

(2·13 m) long and 3 ft (0·91 m) wide, containing the switches, dials, meters etc, for the inter-related systems of guidance and navigation, stabilization and control. During flight the cabin's 210 cu ft (5·95 m³) has a 100% oxygen atmosphere at 5 psi (0·35 kg/cm²); but as a result of the 1967 fire, during ground tests and countdown a less flammable 60/40 oxygen/nitrogen atmosphere of 15 psi (1·05 kg/cm²) is used, and gradually changed after lift-off by the environmental control system, which maintains a comfortable 'shirtsleeve' temperature of 70–75°F (21–24°C). The CM's outer shell consists of stainless steel honeycomb between stainless steel sheets covered on the outside with ablative, or heat-dissipating material. The heat-shield on the base of the cone is 2·75 in. (6·98 cm) thick, and a type of reinforced plastic (phenolic epoxy resin) which turns white hot, chars, and melts away so that the 3000°F (1649°C) re-entry temperatures do not penetrate the surface of the spacecraft. The CM's inner shell is aluminium honeycomb between aluminium alloy sheets separated by a layer of insulation. Food, water, clothing, waste management and other systems are packed into bays which line the walls. There are two hatches, one at the side for entry, and one at the top, or nose, for use when docked with the LM. There are five windows: two side windows, 13 in. sq (83·9 cm²), for observation and photography; two triangular, 8 by 13 in. (20·5 by 33 cm), used for rendezvous and docking; and a hatch window.

Two reaction control engines and the earth landing system are housed in the forward, or nose, compartment; ten reaction control engines, propellant tankage, helium tanks, water tanks, and the CSM umbilical cable, are housed in the aft compartment.

Artist's drawing of Lunar Module preparing to dock with Command and Service Module

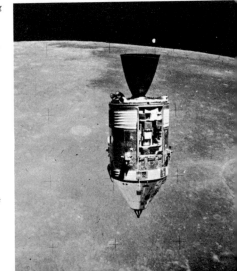

Apollo 15's Command and Service Module showing Scientific Instrument Module Bay

Service Module A relatively simple cylindrical structure 24 ft 9 in. (7·4 m) long, including the 9 ft 9 in. (2·8 m) SPS (Service Propulsion System) nozzle extension. The SM contains the main spacecraft propulsion system and supplies most of the consumables. Total weight rose to 54,063 lb (25,033 kg) for the last three missions, designated 'J', of which 40,593 lb (18,413 kg) was propellant for the 20,500-lb (9300 kg) thrust SP engine. SPS fuel is 50/50 hydrazine and unsymmetrical dimenthyl-hydrazine; oxidizer is nitrogen tetroxide. Since the SM is attached to the CM's heatshield, it cannot be entered by the astronauts, and is jettisoned and burnt up during re-entry. Around the centre section containing the SPS engine and helium tanks, are six pie-shaped sectors: Sector I—oxygen tank 3 and hydrogen tank 3, J-mission SIM bay; Sector II—space radiator, + Y RCS package, SPS oxidizer storage tank; Sector III—space radiator, + Z RCS package, SPS oxidizer storage tank; Sector IV—three fuel cells two oxygen tanks, two hydrogen tanks, auxiliary battery; Sector V—space radiator, SPS fuel sump tank, − Y RCS package; Sector VI—space radiator, SPS fuel storage tank, − Z RCS package.

The Scientific Instrument Module (SIM) was carried for the first time on Apollo 15. Wedge-shaped, 9 ft 5 in. (2·87 m) tall, and 5 ft (1·52 m) across the front, it carried eight experiments, consisting of four spectrometers, panoramic and mapping cameras, laser altimeter, and a 78-lb (35·4-kg) subsatellite for ejection into lunar orbit.

Docking Mechanism This subsystem enables the CM and LM to be connected twice during a normal lunar mission—at the beginning of translunar flight, and when the ascent stage returns from the lunar surface. The probe, with a head housing three capture latches, is mounted in the

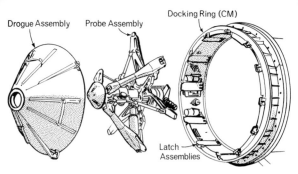

Docking Mechanism

CM docking tunnel and guided into a socket at the bottom of the drogue in the LM docking tunnel. The capture latches hold the two modules together until the probe retraction device (a nitrogen pressure system located in the probe) is operated. Then twelve latches on the CM docking ring automatically operate at contact to form a pressure-tight seal between the modules. When the CM hatch is opened the probe and drogue are dismantled to allow the astronauts to pass through the docking tunnel into the LM.

Lunar Module A two-stage vehicle, 22 ft 11 in. (6·985 m) high and 31 ft (9·45 m) wide diagonally across the landing gear. Its job is to convey two astronauts to and from the CSM in lunar orbit (Plate 3), and to provide living quarters and a base of operations on the moon.

Total weight, with crew, rose from approximately 32,400 lb (14,696 kg) on the early missions, to approx. 36,700 lb (16,647 kg) on the last three 'J' flights, when the LM was carrying the Lunar

Rover as well as extra scientific equipment, batteries and consumables needed for a three-day staytime on the moon. Because the vehicle is operated only in the vacuum of space, aerodynamic symmetry was unnecessary; design was dictated only by the need to provide a descent stage able to serve as a launching platform for an ascent stage containing a pressurized crew cabin. Built by Grumman Aerospace Corp, the LM is designed to operate for 73 hours while separated from the CSM, with a nominal 67 hours spent on the moon.

Ascent Stage The LM's control centre. Constructed of welded aluminium alloy, surrounded by a 3-in. (7·6-cm) thick layer of insulating material; a thin outer skin of aluminium covers the insulation. The cabin provides 235 cu ft (6·65 m³) of space, and is manned by the Commander at the left flight station, and the LM Pilot at right. Even in lunar gravity couches are unnecessary, so armrests and harness-like restraints are used to support the astronauts, giving them sufficient freedom of movement to operate all the controls while in an upright position. There are two inward-opening hatches: the 32-in. (81-cm) diameter docking tunnel hatch at the top; and the 32-in. (81-cm) square forward hatch, located between the astronauts, through which they must back, on hands and knees, to the platform and ladder when descending to the moon. Each crewman has a triangular window, canted down for sideways and forward visibility; a third, docking window, 5 in. by 12 in. (12·7 by 30·5 cm), is directly over the commander's position. The LM has essentially the same subsystems as the CSM, including propulsion, environmental control, communications, reaction control and guidance control.

The ascent propulsion engine develops 3500 lb (1587 kg) thrust, can be fired for a total of 460

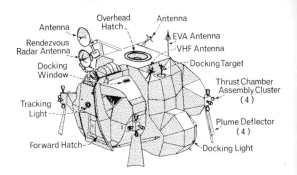

Antenna
Overhead Hatch
Antenna
Rendezvous Radar Antenna
EVA Antenna
VHF Antenna
Docking Window
Docking Target
Thrust Chamber Assembly Cluster (4)
Tracking Light
Plume Deflector (4)
Forward Hatch
Docking Light

LUNAR MODULE - ASCENT STAGE

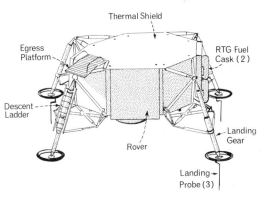

Thermal Shield
Egress Platform
RTG Fuel Cask (2)
Descent Ladder
Rover
Landing Gear
Landing Probe (3)

LUNAR MODULE - DESCENT STAGE

Apollo 16's Lunar Module and Lunar Rover

seconds, and can be shut down and restarted up to thirty-five times (a facility enabling it to replace the main SM engine in an emergency, as on Apollo 13, before lunar touchdown).

Descent Stage The unmanned portion of the LM, representing two-thirds of the total LM weight at earth-launch. Since it has to support the ascent stage for the landing, and later act as its launching platform, it needs a larger engine and more propellant. The gimballed engine provides 9900 lb (4490 kg) max thrust, and a minimum of 1280 lb (580 kg) thrust for the complex 720-second descent manoeuvres from lunar orbit. Its 19,600 lb (8890 kg) of propellant provides 950 seconds of life, with 227 seconds of surface hovering time (compared with 120 seconds on early missions). The Descent Stage's egress platform, or 'porch', is mounted above the forward landing gear strut, which is provided with a ladder of nine steps for descent to the lunar surface. The four landing legs are released explosively; the main struts are filled with crushable aluminium honeycomb to absorb the landing impact; there are also footpads 37 in. (94 cm) in diameter to provide surface support. All the pads except the forward one are fitted with a 68-in. (173-cm) long lunar surface sensing probe which, upon contact with the surface, signals the crew to shut down the descent engine. The landing radar, which controls the descent from a height of 40,000 ft (12,190 m) and speed of 2386 mph (3840 kph), is housed in the Descent Stage. Around the Descent Engine are four equal-sized bays, or quadrants, which, in addition to propellant tanks, contain the MESA, or Modularized Equipment Stowage Assembly, consisting of TV equipment, lunar sample containers, and PLSS, or Portable Life Support Systems; the Lunar Rover Vehicle; and the ALSEP or Apollo Lunar Surface Experiment Package.

Lunar Rover

Earth Weight:	462 lb (209 kg)
Payload:	1080 lb (490 kg)
Length:	122 in. (310 cm)
Max Speed:	8·7 mph (14 kph)
Range:	57 miles (92 km)

The Lunar Roving Vehicle (LRV), built by the Boeing Company, was first used with great success on Apollo 15. This manned spacecraft on wheels had to be built to all the exacting specifications of Apollo hardware, in order to operate in temperatures reaching 250°F (121°C), in a vacuum which ruled out air-cooling. Built and delivered in only 17 months, it carries more than twice its own weight (400 lb (181·5 kg) for each astronaut and his equipment, plus 280 lb (127 kg) of tools, equipment, TV and communications gear, and lunar

Apollo 15's Lunar Module on Hadley Rille

samples). The average family car carries only half its own weight.

Two complete 36-volt battery systems, each of which can power the vehicle on its own, provide a total of 1 hp (1·01 cv), each wheel being individually driven by an electric motor of $\frac{1}{4}$ hp (0·253 cv). It has 78 hours' operational lifetime during the lunar day, and provides the astronauts with a 6-mile (9·65-km) exploration radius from their touchdown point—the limitation being the walk-back distance in the event of a breakdown. Design difficulties were increased by the need to take it to the moon folded into a tight, pie-shaped quadrant of the Lunar Module's descent stage, yet able to unfold itself on the moon, with the astronauts merely pulling sequentially on two nylon operating tapes and then removing a series of release pins. Special 1-G trainers had to be built for earth tests, since the lightweight lunar vehicles could not rest on their own wheels on earth.

A T-shaped hand-controller located on the con-

LUNAR ROVER

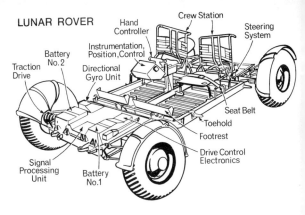

trol and display console between the two seats enables the LRV to be driven by either astronaut. Pushing forwards sets the vehicle in motion; pushing sideways turns it; and pulling backwards applies the brakes. It can climb and descend slopes of 25° and negotiate obstacles and small crevasses—though the seat belts are essential safeguards against the pitching and bouncing experienced when travelling in one-sixth gravity. A dead-reckoning navigation system, set before they drive off and using the sun-angle as a bearing, enables them, no matter how much they twist and turn, to know exactly the direction and distance back to the Lunar Module.

By enabling the astronauts to expend less energy, and thus use their oxygen and cooling water at a slower rate, the LRV doubles the astronauts' stay-time on the moon. Its remotely controlled TV unit also enables Mission Control and the public to watch whatever is being done on the moon, the one drawback being that the umbrella-like antenna can only be unfurled for direct transmissions to earth when the vehicle is stationary so that views of the changing scene during the drives are not possible, though radio contact does continue.

Future Development Many possibilities are being explored. Remote control to enable use of vehicles to continue after the astronauts return is the first and most obvious; another possibility is to send the vehicle, by remote control, to the next landing site, so that a subsequent mission would have to take only fresh batteries instead of a completely new vehicle. Remotely operated versions are inevitable for initial Martian exploration. For future lunar exploration it would be advantageous to convert the Lunar Module itself into a vehicle, so that the astronauts could travel in pressurized conditions.

Lunar Rover unfolding from the Lunar Module

Apollo Missions

Apollo 7   October 1968   The first manned flight of the Command and Service Modules conducted in earth orbit, lasting nearly 11 days (260 hours and 163 orbits).   Walter Schirra, Donn Eisele and Walter Cunningham successfully tested the operational qualities of the space vehicle, measured the performance of the Service Module's main propulsion engine by firing it both automatically and manually on eight occasions; simulated the all-important manoeuvre of extracting the Lunar Module (which was not in fact carried) from the S4B third-stage rocket; and checked out the performance of the heat-shield during re-entry. The astronauts cross-infected each other with severe colds, and there was some irritation in exchanges between them and Mission Control over the requirement for TV transmissions on top of many experiments.   Schirra resigned from NASA soon after the flight.   But it set the pattern of success which carried the Apollo Project right through to its successful moonlandings.

Apollo 8   December 1968   Man's first flight round the moon.   Frank Borman, James Lovell and William Anders were the first men to be launched by the 3000-ton Saturn 5 rocket; NASA's immense courage in electing to make this a lunar flight, with ten orbits round the moon if all was going well, was justified by the complete success of the mission.   The critical lunar orbit insertion (LOI) manoeuvre was achieved by firing the SPS engine behind the moon on December 24; after that, the astronauts spent 20 hours circling the moon, filming and photographing the far side, never before seen by man, as well as obtaining pictures of craters, rills and potential landing sites

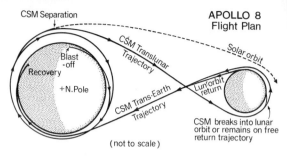

CSM Separation

CSM Translunar Trajectory

Solar orbit

Blast-off

Recovery

+ N.Pole

Lun.orbit return

CSM Trans-Earth Trajectory

CSM breaks into lunar orbit or remains on free return trajectory

(not to scale)

on the nearside. It was at this time that Frank Borman took the famous 'Earthrise' picture, showing the earth rising above the moon's horizon; and, on Christmas morning, stirred the whole world with his reading of the 'In the beginning' passage from Genesis. Splashdown in the Pacific, at the end of the 147-hour mission, was just 11 seconds earlier than the time, computed months before, given in the Flight Plan. The performance of the spacecraft and its rocket engines, and the ability to make precise mid-course corrections when required, provided the final evidence that NASA had the technical equipment and knowledge to land men on the moon. All that was left, was proof that the so-far untried Lunar Module would perform equally well.

Apollo 9 March 1969 The first manned flight with the Lunar Module. James McDivitt, David Scott and Russell Schweickart were launched after a 2-day postponement because of sore throats and nasal congestion. Although the 241-hour flight was confined to earth orbit, separation of the Lunar Module from the Command Module, followed by rendezvous and docking, were carried out in conditions simulating those to be used on the later lunar missions. After the spacecraft was placed in earth orbit, transposition and docking was achieved for the first time—that is, the Command Module separated from the S4B third-stage rocket, turned around, and then nosed in to dock with the Lunar Module and withdraw it from its

Command and Service Module/Lunar Module separating from the S4B third-stage rocket

housing inside the S4B. McDivitt and Schweickart then opened up the docking tunnel, and went through into the LM. With the two vehicles still docked, they test-fired the LM's 10,000-lb (4535-kg) thrust rocket engine. A 2-hour spacewalk by Schweickart was cancelled because he had suffered nausea and vomiting but, with both spacecraft depressurized, he climbed out of the LM, stood on the porch, and tested out the ladder on the LM's landing leg. Scott, standing up with head and shoulders through the open CM hatch, took some memorable pictures of Schweickart's activities (Plate 2), which also provided a satisfactory test of the Extravehicular Activity (EVA) suit in space conditions. On the third day of the flight, with McDivitt and Schweickart again in the

Apollo 9 recovery

LM, and Scott in the CM, the two craft were separated. When 113 miles (182 km) apart the LM's Descent Stage was jettisoned and the Ascent Stage simulating a take-off from the moon, successfully fired its 3500-lb (1590-kg) thrust rocket engine to rendezvous and dock with the CM.

Apollo 10 May 1969 This 8-day (192-hour) flight by Thomas Stafford, Eugene Cernan and John Young was the successful dress rehearsal for the actual moonlanding 2 months later. The first time the complete Apollo spacecraft had orbited the moon, it took man to within 9 miles (14·5 km) of the lunar surface. The whole mission closely followed the Apollo 11 flight plan. The CM remained in a 69-mile (111-km) lunar orbit for nearly thirty-two revolutions, with Young in control, while Stafford and Cernan undocked and made a simulated landing in the LM by twice descending to within 9 miles (14·5 km) of the surface. A moment of great hazard occurred when the LM's Descent Stage was jettisoned after the second close-approach, prior to firing the Ascent Stage engine to rejoin the CM. The Ascent Stage began pitching violently, and it took Stafford about one minute to stabilize it; it was afterwards found that a control switch, omitted from the detailed check list, had been left in the wrong position. The spacecraft successfully docked after 8 hours of separation. Apart from its technical success, this flight was notable because, for the first time, all three astronauts remained in excellent health, both during the flight and afterwards—the first time this had happened on any US manned flight. Another first was colour TV, including a memorable sequence showing the LM firing its attitude control jets as it moved away from the CM prior to its descent towards the lunar surface.

Apollo 11 July 1969 The historic flight by Neil Armstrong, Edwin Aldrin and Michael Collins which culminated in Armstrong and Aldrin becoming the first men to step on the moon. Launch from Cape Kennedy was achieved without any technical problems on Wednesday, July 16. Lunar orbit insertion was successfully achieved on Saturday, July 19, and the spacecraft placed in a 62 by 75 miles (100 by 121 km) orbit; Apollo 8 and 10 experience had shown that by the time the Lunar Module docked with the Command Module after the landing, the magnetic effect of the moon's 'mascons' (mass concentrations) would have gradually changed the orbit into a circular 69 miles (111 km). On Sunday, July 20th, while Collins remained in control of the CM (code-named Columbia) (Plate 5), Armstrong and Aldrin entered the LM (code-named Eagle). The two

39

spacecraft separated on the thirteenth lunar orbit, and Eagle's descent engine was fired behind the moon. As he neared the surface Armstrong decided to take over manual control because the spacecraft was approaching an area in the Sea of Tranquillity strewn with boulders (Plate 5). At 21.17 BST (16.17 EDT) on Sunday, July 21, with less than 2 per cent of descent propellant remaining, more than 500 million people heard Armstrong tell Mission Control: 'Contact light. O.K., engine stop . . . Houston, Tranquillity Base here. The Eagle has landed.' Mission Control replied: 'Roger, Tranquillity. We copy you on the ground. You've got a bunch of guys about to turn blue. We're breathing again. Thanks a lot.' After checking the LM's systems in case an emergency take-off was required, Armstrong and Aldrin elected not to take the 4-hour rest-period scheduled in the flight plan, but to go straight ahead with preparations for their $2\frac{1}{4}$ hours of activity on the moon's surface. The arduous process of donning spacesuits and backpacks, and depressurizing Eagle, took a good deal longer than expected. But finally, Armstrong, on hands and knees, backed carefully out of the spacecraft on to the small platform or porch outside, and descended the 10-ft (3·05-m) ladder attached to the landing leg.

At 03.56 BST on Monday, July 21 (22.56 EDT on Sunday, July 20) he became the first man to step on the moon (Plate 6). As he placed his left foot on the dusty surface, a breathless world could just distinguish his words: 'That's one small step for a man; one giant leap for mankind.' Aldrin joined him on the surface 18 minutes later, and the two astronauts practised the best way of moving about in one-sixth gravity by walking, running and trying out a 'kangaroo hop'. They erected a

Aldrin,
photo-
graphed by
Armstrong,
descending
from Apollo
11's Lunar
Module

Deploying
man's first
scientific
station on
the moon

5-ft (1·52-m) US flag, extended on a wire frame since there is no wind on the moon, saluted it, and received a congratulatory telephone call from President Nixon. A plaque fixed to the space-craft's ladder was unveiled, which read: 'Here men from the planet Earth first set foot upon the Moon, July 1969 A.D. We came in peace for all mankind.' It bore the signatures of the three Apollo 11 astronauts and of President Nixon. Armstrong and Aldrin collected 48 lb (21·75 kg) of rock and soil samples, and placed on the surface of the moon a microdot disc containing messages from most of the world's leaders, as well as a scientific package. This included a laser reflector to enable scientists to measure earth-moon dis-tances by reflecting light beams from it; and a seismometer to measure meteorite impacts and moonquakes. (The seismometer was so sensitive that one of the first signals it transmitted to earth was the impact of the astronauts' heavy boots when they threw them out of the LM to help lighten it before take-off.)

The launching of the Ascent Stage went as planned, and docking was achieved 3½ hours later; there was some vibration at that point, and Collins was heard to say 'all hell broke loose'. After vacuum-cleaning their clothing and equipment, and taking many other precautions to avoid bring-ing back to earth any germs or contamination from the moon, Armstrong and Aldrin transferred to the CM; the LM was jettisoned, the Service Pro-pulsion System was fired behind the moon, and an uneventful return journey achieved. Splash-down was at 17.50 BST on Thursday, July 24 (12.50 EDT), only 30 seconds later than predicted before the start of the 195-hour mission. After donning Biological Isolation Garments, the astronauts were taken by helicopter to the recovery

A close-up of lunar rocks contained in the first Apollo 11 sample return container

ship, where they immediately entered the Mobile Quarantine Facility (a modified trailer), in which they travelled to the Lunar Receiving Laboratory at Houston, Texas. Thus they were kept in isolation—a precaution which proved to be un-necessary—for 21 days after lift-off from the lunar surface, to ensure that no lunar contamination was brought back to earth. During their 8-day mission the Apollo 11 astronauts had travelled 952,700 miles (1,533,225 km).

Apollo 12 November 1969 This second lunar landing mission, by Charles Conrad, Richard Gordon and Alan Bean, was in some ways an even more remarkable flight than Apollo 11. It started sensationally, for as the launch was being made through a rain squall, the Saturn 5 rocket was struck by lightning. The spacecraft's electri-cal system was put out of action for a short time, and Mission Control lost contact briefly. For the first time during a manned launch they were very close to an 'Abort' situation. It was feared that the lightning strike might have damaged the Apollo computer's 'memory' containing the data required for the flight trajectories and the moonlandings. But after being checked during the first 2 hours in earth-parking orbit, all systems were found to be in good order, and astronauts and Mission Control agreed that the flight should continue. Because it was an all-Navy crew, on this occasion when they separated in lunar orbit, the Command and Lunar Modules were code-named Yankee Clipper and Intrepid, with Conrad and Bean making the landing. Because Apollo 11 had touched down 4 miles (6·5 km) beyond the target point, the planners had decided that a series of very small manoeuvres, such as making a water dump, having the LM make a turn so that the CM could inspect it before they moved apart etc., had had a cumulative effect upon the orbit. Extra course corrections were made, and a 'soft' undocking carried out.

The final result, at the end of the 360,000-mile (579,365 km) journey, was the incredible achievement that Intrepid touched down, as planned, only 600 ft (183 m) from the Surveyor 3 spacecraft which had been soft-landed in the Ocean of Storms thirty-one months earlier. Conrad expressed concern later about his landing difficulties, which were increased by a much bigger dust-storm blown up by the descent engine than in the case of Apollo 11; but he did have a safety-margin of 58 seconds of hovering time left compared with Apollo 11's 20 seconds. Conrad and Bean were on the moon for $31\frac{1}{2}$ hours, during which they made two excursions, totalling $7\frac{1}{2}$ hours on the surface. In addition to collecting over 75 lb (34 kg) of rocks and soil samples, they deployed the first ALSEP package—six scientific experiments, with a nuclear-powered battery giving the equipment an operational life of at least a year—and brought back the TV camera and other parts of the Surveyor spacecraft, so that scientists could discover the effects on them of long-term exposure to the solar wind, and of the extreme variations of temperature in vacuum conditions. Conrad became the first man to fall on the moon, but was quickly helped by Bean, and remarked that, despite his cumbersome spacesuit, it was 'no big deal'.

When they had rejoined Yankee Clipper, Intrepid was deliberately crashed on to the lunar surface. The impact 45 miles (72·5 km) from the newly installed seismometer, set the moon 'ringing like a bell', as one geologist put it; the vibrations continued for 51 minutes, an effect completely unanticipated. One theory advanced was that it indicated that the moon was an unstable structure, and that the impact had set off a series of collapses and avalanches. Although the astronauts were again kept in quarantine for 3 weeks after leaving the lunar surface as a precaution against bringing back any possible contamination to earth, it was again found to be unnecessary. Anxiety on the subject was so far relaxed, that on this occasion, nearly 30 lb (13·6 kg) of moonrock samples were distributed to scientists around the world by normal mail services and through diplomatic bags. The only important thing that went wrong throughout this 10-day mission—2 days longer than Apollo 11's—was that Intrepid's colour TV camera failed to work, so that viewers on earth were unable to watch lunar activities.

Apollo 13 April 1970 This third moonlanding attempt was intended to be the first of three landings devoted to geological research on the lunar surface. Instead, an explosion on **board when** the spacecraft was 205,000 miles (329,915 km) from earth all but cost the lives of the crew, and turned the mission into a $3\frac{1}{2}$-day rescue drama surpassing any space-fiction story as tens of thousands of technicians in the US aerospace industry worked to bring the crippled spacecraft safely home. The flight was on the verge of postponement for a week before the launch date because the Commander, Jim Lovell, about to make his fourth spaceflight, Command Pilot Thomas Mattingly and Lunar Pilot Fred Haise, neither of whom had flown before, had been in contact with German measles. Finally it was decided that only Mattingly had no immunity to the disease, and he was replaced with the back-up command pilot, Jack Swigert. Apollo 13 was then launched on time, Mrs Marilyn Lovell declaring that she was not in the least worried because it started on April 13, at 13.13 hours Houston time. Once again it was a less-than-perfect launch; the centre engine amid the cluster of five on the second stage shut down 2 minutes early. The other four burned to depletion to make up the lost thrust, and the third stage, Saturn 4B engine, was automatically fired an extra 10 seconds to put the spacecraft into the required orbit. However, it was decided that the third-stage had plenty of fuel left for the Translunar Injection manoeuvre, and the flight continued. Its progress towards the moon then became so uneventful and routine that public interest in the expedition rapidly waned. But at 55 hours 48 minutes into the flight came the explosion. It was 10 p.m. at Mission Control at Houston, 5 a.m. in Britain; the astronauts had just been congratulated on a routine, but successful TV transmission, and were being advised on how to locate Comet Bennett, when Swigert interrupted sharply: 'Hey,

46

The severely damaged Apollo 13 Service Module
photographed just after it was jettisoned

we've got a problem here.' There had been 'a
pretty large bang', followed by a 'Main B Bus
interval'. (The spacecraft's two major electrical
harnesses are known as A and B, and called 'buses'
by engineers; an 'interval' is a loss of power.)
Only later was it established that the module had
been ripped open by an explosion amid its oxygen
tanks; but 10 minutes after, amid mounting
tension as oxygen readings in the spacecraft and at
Mission Control fell to zero, the crew reported they
could see 'a gas of some sort' venting into space.

Only 80 minutes after the crisis began, journalists
were told that the moonlanding was abandoned,
and that the aim must be to swing the spacecraft
round the moon and use the Lunar Module's
power systems and supplies to bring it back to
earth.   Ten minutes after that Mission Control
told the crew: 'We're starting to think about the
LM lifeboat,' and Swigert replied: 'That's some-
thing we're thinking about too'—the only spoken

acknowledgment at any time of the acute danger they faced. At first it was thought that, even by using the LM as a lifeboat to tow the crippled spacecraft home, only 38 hours of power, water and oxygen were available—about half as much time as would be needed to get the craft home. But as technicians brought their computers and simulators into use, techniques were devised for powering down the systems so that, although it meant increasingly cold and uncomfortable conditions, there would be an ample margin for the return. A major problem arose because the LM's air-conditioning equipment could not extract the poisonous carbon dioxide from the docking tunnel and CM as well; so a do-it-yourself air conditioner was devised on the ground from canisters, space-suit hoses etc., known to be available in the spacecraft, and instructions on how to make it were read up, hour after hour, to the crew. Four firings of the LM's descent engine were carried out in an elaborate series of manoeuvres—accompanied by heart-stopping moments on two occasions when the astronauts, tired and chilled, made mistakes.

The CSM, code-named Odyssey, jettisoned the SM $3\frac{1}{2}$ hours before splashdown, and as it moved away there were exclamations from the astronauts as they saw—and photographed—for the first time the extent of the damage. Finally, the LM, code-named Aquarius, was jettisoned and burned up, one hour before re-entry began. At that time, no one knew for certain whether the explosion had damaged the heat-shield. Evidence that it had not was dramatically provided when the spacecraft, descending on its three parachutes, suddenly filled the TV screens of the world's hundreds of millions of viewers. Apollo 13 ended as a triumphant failure; Lovell, in four flights had completed 715 hours, or nearly 30 days, in space;

NASA pointed out that US manned spacecraft had flown 66 million miles (106·215 million km) without losing an astronaut in space. But it had been a very near thing, and hastened dramatic changes in the whole pattern of future Apollo flights.

Apollo 14 February 1971 The third successful moonlanding expedition, commanded by Alan Shepard, America's first man in space, and the only one of the original seven Mercury astronauts finally to set foot on the moon. Shepard's leadership, together with the immense physical endurance he displayed, proved that criticisms of his selection as Commander at the age of forty-seven, and after recovery from serious ear trouble, were completely unjustified. He had only flown one 15-minute space 'lob', and his companions, CM Pilot Stuart Roosa, and LM Pilot Edgar

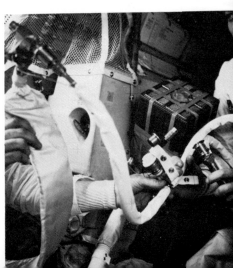

Swigert working on emergency air conditioning and water supply equipment in the Lunar Module 'lifeboat', following the Apollo 13 explosion

Mitchell, had had no previous spaceflight experience at all, so this was easily the least experienced of the Apollo crews. For the first time an Apollo launch was late; the countdown was held for 40 minutes 2 seconds after reaching T − 10 minutes. This was in conformity with new rules following the lightning strike when Apollo 12 was launched. There were again rainclouds and squalls in the Cape area, and the launch was finally made through a hole in the weather, and against a background of distant lightning flashes.

The flight to the moon was accompanied by a series of technical problems, the first and most serious occurring soon after TLI—3 hours 14 minutes after lift-off, and 7200 miles (11,590 km) from earth. At the end of the Transposition and Docking Manoeuvre, Roosa had to make six attempts before he finally succeeded in docking with the Lunar Module. He reported that the probe had been correctly aligned with the hole in the drogue, but instead of the capture latches locking the two craft together, the CM bounced out again. The reasons for this became the subject of anxious discussion between Apollo 14 and Mission Control throughout the flight to the moon, with consideration as to whether it was safe to continue in case the docking manoeuvre proved equally difficult and fuel-consuming when Antares, the LM, returned from the lunar surface. One possibility was considered to be that Roosa carried out his dockings too gently; another that there was some form of 'contamination', such as ice, on the capture latches, since the probe, when dismantled and brought inside the spacecraft, was found to be in perfect order. During the remainder of the flight to the moon, it was possible to cancel the third midcourse correction, and make other fuel economies which finally compensated for the

Apollo Launching Sites on the Moon

A11—Sea of Tranquillity   A12—Ocean of Storms
A14—Fra Mauro site   A15—Hadley Rille
A16—Cayley Plains of the Descartes site
A17—Taurus-Littrow site

extra quantity used during the docking problem. Perhaps the most important experiment made during the outward journey was a study of the optical flashes thought to be due to cosmic rays penetrating the optic nerves, and frequently noticed by astronauts.

For the first time on this mission the CSM was placed in a very low lunar orbit, with an apocynthion of 70 miles (112 km), but low point of only 50,000 ft (15,250 m). This meant that the LM would have more fuel available for hovering and selecting a safe landing point. After separation on the twelfth lunar orbit Roosa took Kitty Hawk, the CSM, up to a 70-mile (112-km) circular orbit. Antares had problems with the abort system and landing radar, but Shepard finally brought the spacecraft to a safe touchdown at 10.18 GMT on February 5. The landing point was only 87 ft (26·5 m) from the target. Although it was 75 seconds late, there were 60 seconds of fuel left, compared with Armstrong's 20 seconds on Apollo 11. When Shepard stepped on the moon 50 minutes late because of trouble with his portable communications system, he commented: 'It's been a long way but we're here.' The first EVA, from depress to repress, lasted 4 hours 44 minutes; it took longer to deploy ALSEP than expected, and Mitchell's 'thumper' device, which should have fired small explosive charges on the surface of the moon to be recorded on the seismometer, was not completely successful; only fourteen of the twenty-one charges went off. Next day, Shepard and Mitchell loaded up the MET, and towing it, set off for the rim of Cone Crater, nearly 1 mile (1·6 km) away and 400 ft (122 m) above the touchdown point. After 2 hours 10 minutes they were 50 minutes behind schedule, and tiring. Shepard's heart rate reached 150, Mitchell's 128; they finally

Despite the fact that they were in grave danger the Apollo 13 crew brought back striking pictures of the moon's surface

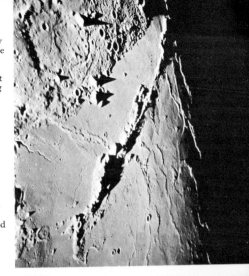

The far side of the moon photographed from Apollo 13's Command Module

turned back short of the rim, and one of the high-lights of the mission, rolling stones down the inside of 125-ft (38-m) deep Cone Crater, had to be abandoned. Shepard made up for it when TV viewers saw him become the first lunar golfer. He produced two golf balls and used the pole of the solar wind experiment to tee off. Their second EVA lasted 4 hours 41 minutes.

There were no docking or other problems on the return flight, the only unusual aspect being that the docking probe was dismantled, stowed and brought back for examination, despite the extra discomfort, and the risk associated with the possibility that it might break away. A record 98 lb (44 kg) of lunar rocks and soil was brought back to earth. The crew were the last lunar landers to have to endure the 3-weeks' quarantine procedure after their return.

Apollo 15 July 1971 The first of three 'J Series' missions, intended to exploit the scientific potential of the Apollo hardware. Lasting 12 days 7 hours, it was 2 days longer than any previous Apollo flight; and such an unqualified success that one estimate was that David Scott, the Commander, Alfred Worden, CMP, and James Irwin, LMP, had brought back as much scientific information as Sir Charles Darwin acquired during his 5-year voyage in *Beagle* in 1831–6. The all-US Air Force crew had in fact named the LM 'Falcon' after their Air Force mascot, and the CSM 'Endeavour' after the ship that carried James Cook on his eighteenth-century scientific voyages. Despite pre-launch problems—the launchpad was struck by lightning eleven times—the countdown itself was flawless, and lift-off took place only 0·008 seconds late. The Saturn 5, developing 7,840,000 lb (3,556,160 kg) thrust, was the most

54

powerful to date; but total lift-off weight, at 6,494,993 lb (2,946,088 kg) was not a record. At 47 tons, (47·75 tonnes), the spacecraft was 2 tons (2·03 tonnes) heavier than previous vehicles sent to the moon—made possible by such measures as reducing fuel reserves and the number of retro-rocket motors on Saturn's first two stages.

Translunar injection, transposition and docking were completed without any of the problems encountered on Apollo 14. A midcourse correction was brought forward 2½ hours after a warning light had indicated a short-circuit in one of the two systems for firing the SPS. For a time the lunar landing was in doubt; but Scott, by manually operating circuit breakers, established that the faulty system could be used for back-up purposes; the mission could therefore continue by using the B, instead of the A system, for SPS firings. Broken glass on an altimeter in the LM, when Scott and Irwin carried out their first inspection, caused some concern throughout the mission—not so much because of the damaged instrument, even though that was for use during the hazardous touchdown, but because of the danger of glass splinters getting into the hatch seals and causing a pressure leak when the LM was finally jettisoned. Alarm over a water leak and signs of flooding were quickly allayed by instructions from Mission Control on how to tighten up a valve. The crew donned their spacesuits, when, a few hours before entering lunar orbit, they blew a 9½ ft by 5 ft (2·90 m by 1·52 m) panel off the SM, to expose eight scientific experiments in the SIMBAY—the bay containing the Scientific Instrument Module. The experiments included mapping and panoramic cameras, and spectrometers, two of them on extendable booms more than 20 ft (6·1 m) long, for measuring the composition of the lunar surface, solar X-ray interaction, and particle emissions while the CSM spent 6 days orbiting the moon. The SIMBAY also carried the 78-lb (35·4-kg) subsatellite, which was successfully ejected following the conclusion of the lunar landing to spend a year studying the moon's mascons and other phenomena.

Falcon made a successful touchdown at 22.16 GMT on July 30, within a few hundred feet of the target, despite much bigger perturbations than expected in Apollo 15's lunar orbit, because it was passing over more mascons than ever before; and despite the fact that separation of Falcon from Endeavour failed at the first attempt because an umbilical, or power line, was not tight. Separation was completed 35 minutes late on the twelfth revolution in time for the planned descent on the fourteenth. Falcon had about 10,000-ft (3050-m) clearance over the top of the 13,000-ft (3960-m) Apennine Mountains as it dropped down to land in a basin near Hadley Rille, a 1200-ft (366-m) chasm named after an eighteenth-century English mathematician (Plate 6). Falcon came to rest with a 10° tilt because one footpad was in a 5-ft (1·52-m) crater—a tilt which caused Scott and Irwin much inconvenience with the swinging hatch door when they were crawling in and out in their cumbersome space suits for three EVAs, which gave them a total of 18 hours 36 minutes on the surface.

They had some difficulty deploying the Lunar Rover, the first lunar motor car, and only on the second EVA, following advice from MC, were they able to get the front-wheel steering to work. Rear-wheel steering was available, however, and though they found that when driving in one-sixth G the vehicle was 'a real bucking bronco', making seat belts essential, Scott and Irwin were

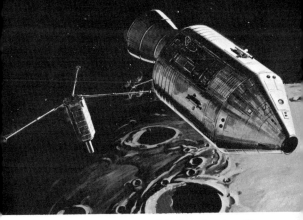

Artist's drawing of the launch of Apollo 15's subsatellite which was to spend a year studying the moon's mascons and other phenomena

delighted with it. When they reached the rim of Hadley Rille, 2½ miles (4 km) from the LM, the world shared their view when they turned on the TV camera mounted on the Rover, and deployed the umbrella-like antenna which transmitted excellent pictures direct to earth. On this first EVA, however, energy and thus oxygen consumption was 17 per cent greater than expected, and MC warned that activities would have to be curtailed. Before it ended, however, the third ALSEP was deployed—an exciting moment for the geologists, confident that, by using the Apollo 12, 14 and 15 ALSEPS, they could finally pinpoint, and even predict, the monthly moonquakes occurring in eleven areas (mostly in the Copernicus region) each time the moon passes at its closest point to earth. MC gave much unavailing advice to Scott, who was watched by millions as he struggled with the percussion drill, leaning on it and panting heavily as he bored holes in the surface, both for bringing back core samples and inserting instruments for measuring the heat flow from the interior. One core, $8\frac{1}{2}$ ft (2·6 m) long, was found to contain fifty-seven separate layers of soil, illustrating 2400 million years of lunar history. The thickness of the layers ranged from $\frac{1}{2}$ in. to 5 in. (1·27 cm to 12·7 cm), each probably representing a different meteorite impact which deposited a new layer of loose top soil.

The second EVA started badly. First there was a 25-lb (11·3-kg) water leak to be mopped up; then Irwin had trouble with air bubbles in his PLSS, and had to get Scott to tape on his communications antenna, which had broken off. But their 7·8-mile (12·5-km) tour of Hadley Mt's foothills, collecting rock samples, proved to be the most rewarding period so far on the moon. Scott excitedly announced that he thought they had

Scott leaning on the percussion drill that gave so much trouble

discovered what they went to find: a 4-inch (10-cm) piece of crystal rock—geologists call it 'anorthosite'—believed to be part of the pristine moon, and quickly dubbed the 'Genesis rock'. (Later, geologists decided it was 4150 million years old, and 150 million years older than any previous sample recovered; but it was *not* as old as the moon itself.) Irwin took a spectacular tumble, and had to be helped up by Scott—who in turn fell during the third EVA. Because they adapted themselves unexpectedly quickly, benefiting both from the Rover and from their improved and more flexible spacesuits, Scott and Irwin used much less energy on the last two EVAs, and they did not have to be shortened as much as anticipated. However, with MC insisting that 100 per cent success had already been achieved, the planned second visit to the rim of Hadley Rille was eliminated. The astronauts finally returned to Falcon after driving a total of 17·4 miles (28 km), and having collected 173 lb (78 kg) of rock samples.

The TV camera on the Rover was carefully set by Scott so that, for the first time, lift-off from the moon could be televised—an astonishing 2 seconds of TV as the screen burst into red and green as the ascent stage rushed upwards, to rejoin Worden, who had had a busy 73 hours alone in Endeavour, conducting the orbital science experiments. There was a 3-hour drama in lunar orbit when jettisoning of Falcon had to be delayed because of doubts as to whether Endeavour's hatch was adequately sealed. The astronauts had to reopen the docking tunnel, check the seals, and delay LM jettison for one orbit; when it was achieved Falcon was in front instead of behind Endeavour, and there was anxiety—and, as Mission Control admitted, 'confusion' there—about how to ensure that the two craft did not collide. Finally the danger was averted, the LM impacted on the moon, and the weary astronauts ignored advice to take sleeping pills at the end of a 20-hour day. Satellite ejection was followed by TEI on the seventy-fourth lunar orbit, then came Worden's 19-minute space-walk, 200,000 miles (321,870 km) from earth, to retrieve two film cassettes from the SIMBAY, according to plan. It was the first spacewalk ever made for a practical, working purpose. Geologists were disappointed when the TV camera on the Rover failed, and they were unable to view the earth being eclipsed by the sun; but interest in this was itself eclipsed by the drama of re-entry, when, for the first time, one of the three 83-ft (25·3-m) parachutes failed to open. The recovery carrier warned the crew to 'stand by for a hard landing'. The crew emerged none the worse for their dramatic splashdown.

The reason for the parachute failure was never conclusively established; most likely causes were either faulty metal links connecting the suspension lines, or residual RCS fuel burning through the shroud lines while being dumped during the final descent. On following missions the links were strengthened and the fuel retained on board.

Apollo 16 April 1972 The fifth moonlanding was a triumphant success, achieved despite the fact that at one period the mission appeared to be in an Apollo 13 situation, with the possibility of an emergency return to earth. It lasted 11 days 1 hour 51 minutes, 24 hours 45 minutes less than the original flight plan. Launch on April 16 was on time, despite a threatened delay until 40 minutes before lift-off because of a gyroscope problem in the Saturn 5 instrument unit. Total lift-off weight was 6,425,000 lb (2,914,000 kg); lift-off

Apollo 16 goes off on time, despite pre-launch problems

thrust was 7,723,726 lb (3,503,405 kg). Spacecraft weight was 107,149 lb (48,602 kg).

The first of about a dozen technical problems which kept John Young, Commander, Thomas (Ken) Mattingly, CMP, and Charles Duke, LMP, troubleshooting for much of the outward journey, occurred $8\frac{1}{2}$ hours after lift-off. Young feared that particles and flakes flowing past the windows might indicate a nitrogen leak; Mission Control instruments showed no such leak, but Young and Duke were ordered to open up the docking tunnel and check out the Lunar Module (Orion) 24 hours earlier than planned. Orion was in good order; and meanwhile, Mattingly, operating the TV camera, enabled Mission Controllers to establish that thermal paint was flaking off some of Orion's panels. After some alarm about possible causes, it was decided it was not serious. Mattingly was soon busy with an electrical signal fault affecting the vital Guidance and Navigation System in the Command Module (Casper); the Massachusetts Institute of Technology worked out a computer programme to ensure that the fault would not abort critical manoeuvres such as lunar orbit and descent orbit insertion. The crew were partially compensated for their extra work; only one of four scheduled midcourse corrections was required.

The major crisis came following undocking on the twelfth lunar orbit. Mattingly, having delivered Orion into a 67 by 12 miles (107 by 19 km) orbit, had to fire Casper's main SPS engine to circularize his orbit and be available for a possible rescue manoeuvre before Orion attempted its lunar landing. He did not do so because of indications of yaw oscillations in the backup system for firing the engine. With only minutes to go before starting their final descent, Young and Duke were given the first 'space wave off': both craft were ordered to continue orbiting, and to reduce the gap between them, ready to re-dock, while Mission Control studied the problem, and even talked of an Apollo 13 situation, in which the lunar module might once again have to be used as a liferaft and tow home a disabled Command Module. But on the fifteenth orbit Christopher Kraft, Houston's director, gave a 'go' for landing. Mattingly, using the primary system, successfully fired the SPS engine to circularize his orbit; Young and Duke finally settled 6 hours late on the Cayley Plains of the Descartes landing site at 0223 GMT on April 21, 656 ft (200 m) north and 492 ft (150 m) west of their Double Spot landmark. They were only about 10 ft (3 m) from a crater some 25 ft (7·6 m) deep.

Young had become the first man to orbit the moon twice; he and Duke, the ninth and tenth men on the moon, became the most enthusiastic, most vocal, and most efficient of them all; but after their exhausting experience during the approach, they elected to sleep before making the first EVA. Duke, who had complained earlier about his spacesuit being tight, had no difficulty at the start of the first 7-hour 11-minute EVA. The one major disappointment on the moon was that Young backed into, and broke, the cable connecting the top-priority heatflow experiment to the ALSEP power plant, thus destroying the scientists' hopes of checking the discovery by the Apollo 15 crew that heat was flowing from the lunar interior at a far higher rate than expected. But an improved drill this time enabled Duke to obtain 10-ft (3-m) deep core samples without difficulty. Although at a later stage a rear mudguard (fender) fell off the Lunar Rover, causing the astronauts to be sprayed with soil, it performed well, and at the end of EVA 1, Young, filmed by Duke, carried out the planned 'Grand Prix'. He drove it flat out in circles, skidding it to test wheel grip.

EVA 2 dramatically demonstrated the improved quality of TV pictures; when they had driven 2·6 miles (4·1 km) to Stone Mountain, and 760 ft (232 m) up it, the astronauts could be seen working on ridges and crater rims; geologists on the ground used the TV camera completely independently to look around, and zoom in for close-ups of boulders and other features of interest. The geologists were embarrassed by the lack of evidence of volcanic activity.

EVA 3 was shortened to 5 hours 40 minutes to meet the revised lift-off time; it was devoted to exploring North Ray crater, a large depression 3 miles (4·9 km) from Orion. Before re-entering for the last time, Young and Duke competed to

Duke skidding the Lunar Rover during the Apollo 16 'Grand Prix'

John Young walks away after deploying Apollo 16's ALSEP. Lunar surface drill is *right, centre*

see how high they could jump; Duke, who had tumbled many times, fell heavily backwards, causing much concern at MC, but escaped with no personal or spacesuit damage. Ascent and docking went perfectly; but a wrongly positioned switch caused Orion to start tumbling immediately after undocking. Casper had to make a hurried evasive manoeuvre, and Orion had to be left in lunar orbit instead of being impacted on to the lunar surface. In view of the slightly suspect backup system for firing the SPS, the journey home started as soon as the 90-lb (40-kg) subsatellite had been ejected. (The planned plane change, to place the subsatellite in a more suitable inclination, was cancelled, to avoid another SPS burn.) The highlight of a relatively quiet return flight was Mattingly's EVA 170,000 miles (274,000 km) from earth to recover film cassettes from the SM SIMBAY. All three parachutes deployed perfectly after re-entry, and were recovered for examination because of the Apollo 15 failure. The CM, however, turned upside-down in the water until the buoyancy balloons were released. Residual propellants were retained on board for the first time, because of fears that they may have damaged Apollo 15's parachute lines; but the wisdom of venting them on previous missions was demonstrated when these propellants exploded after the spacecraft had been taken back to Houston, causing damage and injuring a number of technicians.

A special diet of food and orange-juice laced with potassium enabled this crew to become the first to survive a mission without any irregular

Young with geological hammer collecting samples on crater rim. *Centre* is Gnomon used to establish scale in photographs like this

heartbeats, and removed doubts about the practicability of fifty-six-day flights planned for Skylab. Young and Duke clocked up record totals of 20 hours 40 minutes on the lunar surface; drove 16 miles (27 km); and brought back an estimated 215 lb (97·5 kg) of lunar samples.

Apollo 17 December 1972 Project Apollo is due to start at about 9.38 p.m. EST on December 6 (3.38 a.m. GMT, December 7). It will be the first time America has launched men at night, and therefore a suitably spectacular finale. For the first time the crew includes a geologist-astronaut, Dr Harrison (Jack) Schmitt, as LMP. The Commander, Eugene Cernan, is making his third flight; Ronald Evans, CMP, and Schmitt have not flown before. The Taurus-Littrow landing site was chosen for its combination of the Taurus mountains to the north and the lowland valleys, including the 20-mile (32-km) wide Littrow crater, to the north-west. Cernan and Schmitt must land in a 7-mile (11-km) wide valley between lunar mountains 4500 ft (1372 m) high on one side and 7000 ft (2134 m) on the other. A low range of hills runs through the valley, which is strewn with volcanic ash and some craters which may have been formed by volcanoes—features first reported by Al Worden while orbiting as Apollo 15's CMP. The site is considered to be the most likely one within reach of this mission to fill in the gaps in knowledge on the origins and formation of the moon obtained by earlier flights. The dark material flooding the landing site may have been thrown up from deep inside the moon comparatively recently; another target sampling site is a huge landslide, which has resulted in debris falling into the valley from high up on the 7000-ft mountains.

67

High priority will be given to a renewed heat-flow experiment; new experiments will seek to measure the physical properties of the lunar interior to a depth of 1 km, and to establish whether there is any subsurface water to a depth of $1\frac{1}{2}$ km. Astronaut pressures for a much more adventurous finale to the Apollo programme, such as a landing in the rugged Crater Tycho area in the deep south, or even on the far side of the moon, were ruled out by NASA because of the extra fuel required to achieve the necessary orbit for a Tycho landing, and because of the problem of intermittent communications conducted through the CSM in the case of a farside landing.

Summary and Conclusions The total costs of Project Apollo, including the development and production of all the hardware, the facilities, tracking networks and launch costs, are estimated at £10,000 million ($24 billion). Whatever the results of the final missions, the achievement of man's first moonlanding, the technical mastery of space travel, and the final scientific expeditions to the lunar surface, must inevitably rank as the most significant step so far in the development of man.

Nevertheless, NASA themselves admit that, 'since the Apollo landing sites were highly restricted by operational and safety limitations, most of the moon's surface, containing many intensely interesting and mysterious features, remain untouched, awaiting mankind's next series of contacts.' Since the US has no plans for further manned missions to the moon until the 1980s, by which time the Space Shuttle should have been developed, it seems likely that manned inspection of these mysterious lunar features will be taken over during the next decade by Soviet explorers. But Project Apollo should enable the US, if it so desires, to be the first nation to establish permanent manned bases on Space Station Moon. They now have first-hand experience of the fact that the lunar material is firm enough to support structures, and yet can be easily dug for the construction of shelters and protective barriers. The examination of parts of the Surveyor 3 spacecraft, recovered by the Apollo 12 crew after Surveyor had been on the moon's surface for 31 months, has established that there should be no unexpected difficulties with permanent installations set up on the moon. Most important of all, the total of up to 800 lb (363 kg) of lunar soil and rock samples, expected to be amassed by the conclusion of Apollo 17, should enable techniques to be developed for making lunar bases self-supporting. By the end of Apollo 14, it had already been established that both water and oxygen could be produced from the lunar soil, since it contains a high percentage of iron oxide. By using a solar furnace and introducing hydrogen, 14 lb (6·35 kg) of water could be produced from 100 lb (45·4 kg) of iron oxide, and the water then separated into oxygen and hydrogen. Thus life could be supported and fuel provided for space vehicles.

GEMINI—
US MANNED SPACECRAFT
Gemini 12 Figures

Launch Weight:	8360 lb (3792 kg)
Length:	18·4 ft (5·60 m)
Base Diameter:	10 ft (3·05 m)
Height:	19 ft (5·80 m)
Insertion Velocity:	17,536 mph (28,221 kph)
Apogee:	173·6 m (279·4 km)
Perigee:	100 m (161 km)
Period of Revolution:	95 mins

History Successor to Mercury, the Gemini spacecraft was twice the weight, and drew heavily on proven Mercury technology. But it was also far more advanced, complex and versatile, and achieved far more than its original aim, which was to bridge the gap in US manned spaceflight between the end of Project Mercury and the start of Project Apollo. As it was, 2 years elapsed between the last Mercury mission and the first Gemini flight; but once the programme did begin, ten missions, Gemini 3–12, were flown at a breathtaking rate of five a year in 1965–6. Despite some technical problems, invaluable experience was obtained in long-duration flight, rendezvous, docking, use of target-vehicle propulsion for orbital manoeuvres, extravehicular activity (EVA, or space-walking), and guided re-entry. In those 2 years, the US put ten men into space, while the Soviet Union achieved only the two-man Voshkod 2 flight. It was unfortunate for America that that flight was the occasion of Leonov's historic first spacewalk—12 minutes which enabled Russia to retain her international prestige as leader of the space race despite the solid American progress. It also completely obscured the fact that Gemini 8 achieved the first space docking, and established the technical lead that finally enabled the US to land the first men on the moon years ahead of Russia.

Spacecraft Description Unlike Mercury, which had nearly all its systems inside the pressure shell, Gemini has most systems' components located in unpressurized equipment bays. This enables launch crews to remove a hatch, and quickly replace any malfunctioning component, as well as making possible a larger crew compartment—55 cu ft (1·558 m³) of accessible volume. Titanium and magnesium are the principal metals used. The spacecraft consists of two main parts, a re-entry module and an adapter module.

Re-entry Module This includes the crew compartment, with side-by-side seating and emergency ejection seats, replacing the Mercury escape tower. (These can be used not only for escape during a launchpad or lift-off emergency, but for a

Gemini 7 taken through the hatch window of Gemini 6 during the rendezvous manoeuvre

final parachute descent if the spacecraft is descending over land, instead of sea.) Also part of the re-entry module are electronics for guidance, instrumentation and communications; two hatches; heat shield; a re-entry control section, with propellant tanks and sixteen thrusters in two independent systems for re-entry control; and a rendezvous and recovery section, with radar, drogue and main parachutes.

Adapter Module With a maximum diameter of 120 in. (3·05 m), this provides the bays for rendezvous systems and equipment, and the mating structure between re-entry module and launch vehicle. It contains the life support and electrical systems, with two fuel cells providing the primary source of power from launch to re-entry. Gemini was the first spacecraft to use fuel cells (a British development) providing electrical power through the chemical reaction of oxygen and hydrogen, with each cell producing a pint (0·57 litre) of drinking water per hour as a by-product.

Attitude and Manoeuvring System This consists of sixteen thrusters mounted in fixed positions on the Adapter Module. The desired amount of impulse or control is obtained by varying the firing time: eight of the engines provide 25 lb (11·3 kg) thrust; two produce 85 lb (38·6 kg); and six produce 100 lb (45·4 kg) thrust. An illustration of the amount of attitude and manoeuvring control required is that during the relatively short Gemini 6 flight more than 35,000 individual thruster firings occurred.

Gemini Flights

The first two Gemini flights were unmanned tests; the first seven in the series, launched by Titan 2 rockets, were code-named 'GT'; the last five, which involved dual launches, with an Agena rocket being placed in orbit to act as a docking target, were code-named 'GTA' launches.

Gemini 3 First manned flight in the series—a three-orbit mission in March 1965. The first computer was carried into space—50 lb (22·7 kg) of miniaturized equipment capable of making 7000 calculations a second. It was a feature hardly noticed at the time; but it had much greater long-term significance than Leonov's spacewalk. With it Grissom, the Commander, was able to compute the thrust needed to change his orbit. From that time, instead of being carried helplessly round and round the world on a fixed orbital path, man had a genuine ability to 'fly' in space.

Gemini 4 On this flight, 3 months later, the computer failed and Jim McDivitt had to make a manual re-entry. He was 1 second late punching the button to fire the retro-rockets, and the space-craft came down 40 miles (64·4 km) off course. All the same, it was a triumphant flight, lasting 4 days —the first US long-duration flight. During it, 3 months after Leonov had first tried it, Ed White

Gemini space food

ventured outside. His spacewalk lasted 21 minutes; experimentally, and somewhat uncertainly, he manoeuvred himself around with an oxygen-powered space gun (Plate 2). The colour pictures brought back have been used on books, magazines and newspapers all over the world ever since to symbolize man in space. ('See Voskhod Spacecraft—History'.)

Gemini 5 An 8-day flight by Cooper and Conrad in August 1965, which proved that men could withstand weightlessness for as long as it would take to fly to the moon, visit it briefly and return.

Gemini 7 A 14-day flight in December 1965 by Borman and Lovell, which held the long-duration record for 5 years until finally overtaken by Russia's 18-day Soyuz 9 flight in 1970.

Gemini 6 This mission, postponed in October 1965, at T −42 minutes, when the Agena rocket intended as a rendezvous target was lost, used Gemini 7 as the target instead. On Gemini 7's eleventh day in orbit Schirra and Stafford went up to join them, and Schirra brought his craft within 6 ft (1·83 m) of Gemini 7—the first real 'rendezvous' of men in space (Plate 3). The 1-day Gemini 6 flight should have taken place 3 days earlier, but once again had to be postponed when a fault occurred in the launch vehicle 2·2 seconds before lift-off.

Gemini 8 On March 16, 1966 Armstrong and Scott achieved the first space docking, by linking their spacecraft with an Agena target rocket, placed in orbit 24 hours earlier. Shortly afterwards the linked vehicles began tumbling and spinning, out of control, as a result of a jammed

Gemini 8's
Agena target

thruster.   The astronauts escaped only by firing
their retro-rockets, and had to return to earth 2
days early.   They survived man's first space
emergency, and the drama completely obscured
the fact that it was on this mission that the US
overtook the Soviet Union's long-held lead in
space technology.

Gemini 9   The unlucky flight.   Its original
crew, See and Bassett, was killed when trying to
land a jet fighter in bad weather at the McDonnell
works at St Louis, crashing through the factory
roof only yards from the spacecraft.   The mission
was finally flown by the backup crew, Stafford and
Cernan, in June 1966, after two delays: first because the Agena target was lost; then, when a
second was launched, Gemini 9 missed its window.
But the flight, when it did take place, included a 2-
hour spacewalk by Cernan, and for the first time
splashdown was less than half a mile (0·80 km)
from the recovery ship.

Gemini 10 A 3-day mission in July 1966 during which Young and Collins first docked with an Agena target rocket, then fired its engine to boost themselves into a 475 m (761 km) orbit. There they separated and rendezvoused with Gemini 8's Agena target, which had been left in a 'parking orbit' for this purpose. Collins spacewalked across to it and retrieved a dust-collecting device from the outside of the Agena.

Gemini 11 Conrad and Gordon, in September 1966, made a direct-ascent rendezvous and docked with an Agena target while still on their first revolution. During a 44-minute spacewalk, Gordon connected the two craft with a 100-ft (30-m) tether, and when undocked, Gemini's thrusters were used to put the craft into a cartwheel motion —the first experiment in creating artificial gravity even though it was only 0·00015 of normal earth gravity. Automatic, computer-steered re-entry was achieved for the first time.

Gemini 12 The last mission, flown by Lovell and Aldrin in November 1966, included three spacewalks totalling $5\frac{1}{2}$ hours by Aldrin—the first fully successful attempts at EVA. Once again the spacecraft docked with an Agena target, artificial-gravity experiments were made, and a fully automatic re-entry performed.

Summary When Project Gemini ended nine dockings with a target had been achieved and seven different ways of doing it had been worked out. The ten missions were completed without the loss of a single life. The total cost was £534,750,000 or $1,283,400,000.

MERCURY—US MANNED SPACECRAFT

Launch Weight:	4265 lb (1935 kg)	MA 6
Orbit Weight:	2987 lb (1355 kg)	Weights
Retrofire Weight:	2970 lb (1347 kg)	
Splashdown Weight:	2493 lb (1131 kg)	
Recovery Weight:	2422 lb (1099 kg)	
Height:	9·5 ft (2·90 m)	
Base Diameter:	6 ft 2½ in. (1·89 m)	
Escape Tower:	17 ft (5·18 m)	

History Project Mercury, the first US manned spaceflight programme, was initiated in November 1958; its object was to establish, with a one-man vehicle, that men could be sent into space and returned safely to earth. Six flights started with Alan Shepard's suborbital of 15 minutes on May 5, 1961, and ended on May 16, 1963, with Gordon Cooper's twenty-two-orbit mission lasting 34¼ hours. Redstone rockets were used for the suborbital, and Atlas rockets for the orbital missions.

Spacecraft Details The bell-shape was determined primarily because of heating conditions during re-entry; limited launch capability at that time—Redstone's thrust was 78,000 lb (35,380 kg) —necessitated a design just large enough to contain the astronaut and essential equipment: hence, with some justification, it is usually referred to as 'the capsule'. The main conical portion contains the pilot, the life-support system, the electrical power system, and the systems' controls and displays. The cylindrical section contains the main and reserve parachute landing system. The topmost (antenna canister) section contains RF transmission and reception equipment, and the drogue parachute. The cabin section consists of an inner pressure vessel and an outer heat-resisting skin, fashioned from titanium and nickel alloy, with ceramic fibre insulation. Front and rear pressure bulkheads are fabricated from titanium, the rear bulkhead supporting, among other things, the astronaut's individual, form-fitting couch. This is designed so that the gravity forces during launch and re-entry are evenly distributed throughout his body.

Heat Shield For the suborbital missions, blunt end construction was of beryllium; for the four orbital missions, an ablative shield mixture of glass fibres and resins, so that during re-entry (max temp. 3000°F, 1649°C), the resin vapourized and boiled off at low temperatures into the hot boundary layer of air. The upper part of the spacecraft (max temp. 1300°F, 718°C) was protected with an insulated double-wall structure. The heat shield is detachable, so that following re-entry and main parachute deployment, the shield can be dropped 4 ft (1·22 m) below the spacecraft, pulling out with it a circular collar of rubberized glass fibre. The heat shield thus strikes the water first, absorbing the initial shock; the collar acts as a cushion, and then, as it fills with water, serves as a sea 'anchor'.

Control System Originally designed so that, in normal circumstances, control should be exercised entirely from the ground, though with the astronaut able to take over manually in the event of failure of such things as retro-rocket firings, life-support systems, air conditioning, communications and attitude control. A red light on the panel indicates failure of the retro-rockets to fire at the correct time. Attitude control is by two independent thruster systems: 1) Twelve H_2O_2 pitch, yaw and roll thrust nozzles which can be operated automatically, or by ground control or by the astronaut.

ATLAS–MERCURY

Escape Tower

Spacecraft

Re-entry Vehicle Adapter

Liquid Oxygen Tank

Fuel Tank

Launcher Fitting

28 ft.

95 ft. 4 ins.

2 ft. 3 ins.

16 ft.

MERCURY SPACECRAFT
To Larger Scale

Escape Tower

Antenna Housing

Recovery Compartment (Parachutes)

Crew Compartment

Heat Shield

Retrograde Package

In the last mode, 'Fly-by-Wire', the astronaut operates the nozzles by movements of his control stick. 2) Six more nozzles in a separate system, used either mechanically or manually. The eighteen nozzles provide thrust from 1–24 lb (0·45–10·9 kg). The astronaut is provided with an instrument panel periscope for observing the earth's surface, to provide a datum for manual control of the spacecraft's attitude with the reaction jets in case of failure of the automatic system; it uses infra-red horizon scanners.

Communications Duplicated systems for voice, radar, command, and telemetry links.

Life Support: Two control circuits for cabin and suit, providing breathing, ventilation and pressurization, with 100% oxygen atmosphere at 5·1 psi (0·36 kg/cm²).

Escape Tower Operated by one solid-propellant rocket of 50,000 lb (22,680 kg) thrust with three canted nozzles, and able to lift the spacecraft clear of the rocket for recovery even from ground level. Separation of tower in normal operation is by three solid-propellant rockets of 350 lb (159 kg) thrust.

Retro-rockets Three ripple-firing (which means that burn times overlap) solid propellant rockets of 1160 lb (526 kg) thrust, operated with the spacecraft oriented at 34°, heat-shield forward and up. They reduce the orbital speed of about 17,500 mph (28,165 kph) by 350 mph (563 kph), thus starting re-entry.

Parachutes One 6-ft (1·83-m) diameter conical ribbon-type drogue housed in antenna cone and opening at 21,000 ft (6400 m); 63-ft (19·2-m) diameter ringsail main, and reserve, opening at 10,000 ft (3050 m), and stowed in spacecraft's cylindrical neck.

Mercury Flights

The first research and development spacecraft was successfully launched by an Atlas-D rocket on September 9, 1959. Various test-firings of boosters, spacecraft and escape systems followed. The main flights were named and numbered according to the booster employed: Mercury-Atlas 1, a ballistic trajectory test of an unmanned spacecraft in July 1960, was unsuccessful because of rocket failure. Mercury-Redstone 1, in November 1960, also failed because the rocket engines cut out after it had lifted one inch (2·54 cm) above the pad, but was successfully repeated 1 month later. Mercury-Redstone 2 (MR 2) is described below; Mercury-Atlas 2–4, further unmanned tests, were completed by September 1961. Details of chimpanzee and manned flights are given below.

MR 2 On January 31, 1961 Ham, the chimpanzee, was launched from Cape Canaveral, and endured 17 G during lift-off, and loss of cabin pressure; after being weightless for 6 minutes, he again endured heavy re-entry loads. Owing to a series of malfunctions, the capsule reached a speed of 5800 mph (9335 kph) instead of 4400 mph (7080 kph), and a height of 157 miles (253 km) instead of 115 miles (185 km). It landed 132 miles (212·4 km) from the target, and was shipping water and submerging by the time helicopters recovered it, partially tearing off the heatshield as they did so. But Ham worked throughout the flight, was little the worse, and accepted an apple with a large grin on being released. His survival in spite of the setbacks gave the seven Mercury astronauts confidence in the system.

MR 3 On May 5, 1961 at 9.34 am Alan Shepard,

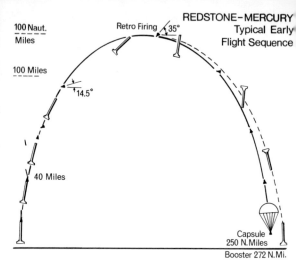

REDSTONE—MERCURY
Typical Early
Flight Sequence

100 Naut.
Miles

100 Miles

Retro Firing 35°

14.5°

40 Miles

Capsule
250 N. Miles

Booster 272 N. Mi.

37, became the first US man in space in Freedom 7, with a flight lasting 15 minutes 22 seconds. He was weightless one-third of that time, rose to an altitude of 116·5 miles (187·5 km), attained a maximum speed of 5180 mph (8336·2 km) and landed 302 miles (486·0 km) downrange from the Cape. He experienced 6 G during acceleration, and 11 G on re-entry. Recovery operations went perfectly, the spacecraft was undamaged, and Shepard exuberant.

MR 4 On July 21, 1961 'Gus' Grissom, in Liberty Bell 7, admitted that he was 'a bit scared' at lift-off on the second suborbital mission. He was weightless for 5 minutes 18 seconds of the 16 minutes 37 seconds flight, reached a height of 118·3 miles (190·4 km) and endured 11·1 G on re-entry. Recovery did not go according to plan. The new explosive hatch-cover, incorporated for quick rescue of an injured astronaut, blew off after Grissom had removed the pin from the detonator and was awaiting arrival of the helicopter. Liberty Bell began shipping water, so Grissom hurriedly climbed out and swam away. Recovery efforts failed to prevent the spacecraft sinking in 15,000 ft (4570 m); but a rather waterlogged Grissom was hauled into the helicopter none the worse for his experience. The planned MR 5 and 6 flights were cancelled as unnecessary.

MA 5 On November 29, 1961 chimpanzee Enos went into a nearly perfect orbit of 99 by 147 miles (159 by 237 km), after holds totalling 2 hours 38 minutes. Because of environmental control system troubles, Enos' temperature rose to over 100°F (37.8°C), and re-entry was ordered after only two instead of the planned three orbits. But recovery was perfectly executed after a 3-hour 16-minute flight, during which Enos was weightless for 2 hours 41 minutes. He survived over 7 G during acceleration and 7.8 G during re-entry, and seventy-nine undeserved electric shocks when one of the control levers, which he had been trained to operate to avoid shocks, malfunctioned. Nevertheless Enos continued to operate it; and it was decided that, had an astronaut been aboard, he would have been able to correct the ECS problem.

MA 6 On February 20, 1962, after delays and postponements starting on January 23, John Glenn, 41, 3 hours 44 minutes after entering Friendship 7, and patiently enduring a whole series of last minute problems, became the first US man in orbit—11 months after Gagarin. His apogee was 162.2 miles (261 km), perigee 100 miles (161 km),

83

and velocity 17,544 mph (28,234·3 kph); he was weightless for 4 hours 48 minutes of the 4-hour 55-minute flight. Glenn was intrigued, as was the world below, with the fact that, at sunrise every 45 minutes, his spacecraft was surrounded with brilliant specks—'fireflies' he dubbed them—which disappeared in bright sunlight. A sticking valve in a reaction jet, similar to the trouble that had necessitated termination of MA 5 after two orbits, enabled Glenn to demonstrate that an astronaut could overcome such problems. Then a telemetry fault started a major alarm by indicating that the heatshield was no longer locked in position. A difference of opinion on the ground was resolved with the decision to order Glenn not to jettison the retropack before re-entry; it was hoped this would help retain the heatshield in position. As the retropack burned and broke away, Glenn commented: 'That's a real fireball outside'; and later, as chunks flew past his window, he thought his heatshield was disintegrating. But all was well, and Friendship 7 splashed into the Atlantic 40 miles (64·5 km) short of the predicted area; retrofire calculations had not taken into account the spacecraft's weight loss in consumables. Glenn's only injury was knuckle abrasions when he punched the detonator to blow off the hatch after Friendship 7, with Glenn still inside, had been hoisted on to the deck of the recovery destroyer. Recovery forces totalled 24 ships, 126 aircraft, and 26,000 personnel.

MA 7 On May 24, 1962 Scott Carpenter, in Aurora 7, became the second US man in orbit, following a near-perfect countdown, delayed only by three 15-minute weather holds. The successful Glenn flight led to a decision to make this a **scientific**, rather than an engineering flight.

Carpenter carried out a number of experiments, including the release of a multi-coloured balloon on a 100-ft (30·5-m) nylon line, which failed to inflate correctly; he manoeuvred the spacecraft with such enthusiasm, yawing and rolling and discovering that even when flying head-down to earth he was not disorientated, that by the end of two orbits he had used more than half his fuel. When eating, he discovered that weightless crumbs floating in the spacecraft could be a danger to breathing. Problems during preparations for re-entry, and a hurried switch to fly-by-wire control with the result that he forgot to switch off the manual system, resulted in the spacecraft running out of both manual and automatic fuel; the retro-rockets, which Carpenter operated by pushing a button, fired 3 seconds late; other errors contributed to Aurora 7 overshooting the Atlantic splashdown target by 250 miles (420 km). It was 50 minutes after splashdown before it was established that Carpenter had struggled out of the

Miss Sam,
rhesus monkey
used in early
Mercury tests

spacecraft and was safe in his liferaft. He had been weightless for 4 hours 39 minutes out of 4 hours 56 minutes, had reached a velocity of 17,549 mph (28,241·6 kph) and endured 7·8 G.

MA 8 On October 3, 1962 with Walter Schirra in Sigma 7, this was 'the textbook flight'. Apart from an alarming clockwise roll 10 seconds after lift-off, and temperature problems with his space-suit which had Ground Control discussing pos-sible termination after the first orbit until Schirra succeeded in adjusting it, the six-orbit flight—double that of previous missions—went well from beginning to end. Schirra carried out disorienta-tion tests, and practised conserving his fuel so effectively, that a final, one-day MA 9 mission at last became possible. He splashed down only 4·5 miles (7·24 km) from the recovery ship in full view of the crew and waiting newsmen. He had been weightless for 8 hours 56 minutes of the 9-hour 13-minute flight, and post-medical checks showed unusual symptoms. His heartbeat, 70 when lying down, rose to 100 when standing; and his legs and feet turned reddish-purple, indicating pooling of blood in the legs. But the condition lasted only 6 hours, and next day his heart and blood pressure were normal.

MA 9 On May 15, 1963 Gordon Cooper in Faith 7 started a twenty-two-orbit flight which lasted 34 hours 20 minutes, of which, as usual, all but 17 minutes was spent in weightlessness. He described the spacecraft as a 'flying camera' since it had a slow-scan TV Monitor and various other photographic systems. Extra equipment, fuel, oxygen, water etc., for the longest Mercury mis-sion had raised the in-orbit weight to 3026 lb (1373 kg). Experiments included the release of a

6-in. (15·24 cm) sphere containing a flashing light which he was able to spot—a significant step towards rendezvous manoeuvres. Cooper also carried out experiments aimed at developing a guidance and navigation system for Apollo; and caused scientific scepticism and interest by claiming to see individual houses and smoke from their chimneys while passing over Tibet at an altitude of 100 miles (161 km). On the nineteenth orbit a warning light suggested the craft was decelerating and beginning to re-enter; by the twenty-first his automatic stabilization and control system had short-circuited and the carbon dioxide level was rising in the spacecraft. Despite his dry comment, 'Things are beginning to stack up a little', Cooper manually conducted his re-entry so efficiently he splashed down only 4 miles (6·44 km) from the recovery carrier.

Performance Summary A 3-day MA 10 mission desired by the astronauts was finally turned down. So Project Mercury lasted 55 months from authorization. While John Glenn's first orbital flight was 22 months later than originally targeted, the final one-day flight took place only 40 months after the decision to make it. Only one of the seven Mercury astronauts got no flight; that was Donald Slayton, originally selected for MA 7; he was found to have an erratic heart rate, and was replaced by Schirra. The total cost of Project Mercury was £163,375,000 ($392,100,000).

SALYUT—USSR ORBITAL
SCIENTIFIC STATION

Weight (with docked Soyuz):	50,000 lb (22,650 **kg**)
Weight (without Soyuz):	36,000 lb (16,300 **kg**)
Length:	65 ft (20 m)
Length (? incl. Soyuz):	65 ft (20 m)
Length (? excl. Soyuz):	33 ft (10 m)
Internal Vol:	3500 cu ft (100 m³)
Internal Temp:	63°F (17°C)—Soyuz
	11 72°F (22°C)

History Salyut, launched on April 19, 1971, remained in near-earth orbits for nearly 6 months, until a final re-entry manoeuvre caused it to burn up on October 11, 1971, after about 2800 orbits. Moscow described the mission as an important step towards the creation of long-term orbital scientific stations in near-terrestrial space; earlier references suggested that Salyut would be followed by similar experimental stations, remaining in orbit for up to a year. The initial orbit of 124 by 138 miles (200 by 222 km) was so low that natural decay would have resulted in re-entry in just over a week. Constant corrections were necessary during Salyut 1's 6 months' life to avoid re-entry. The reasons for the low orbit have never been explained; one possibility is that most of the experiments were connected with earth-observations; another is that the weight of the three-man Soyuz 10 and 11 craft which docked with it had risen near the limit of existing Soviet rocket launch capability.

Following the death of the Soyuz 11 crew after leaving Salyut on June 29, temperature and atmospheric pressures were maintained at life-supporting levels, and the orbit was raised by **posigrade** firings each time the decaying orbit

Artist's drawing of Soyuz 11 docking with Salyut

neared the re-entry point. It seems likely that at least one more Soyuz docking was intended, with a further, and probably even longer period being spent aboard Salyut by the crew. Presumably the inquiry into the Soyuz failure, and subsequent modifications to the spacecraft, were not completed in time; on October 11, when the orbit had once again dropped to 113 by 110 miles (182 by 177 km), Salyut's engine was retrofired, so that it was destroyed as it re-entered the atmosphere. Destruction by natural decay was not permitted because of the risk that parts of the station might have penetrated the atmosphere and fallen on populated areas. Inevitably scientific packages and the results of many experiments were lost. The official government announcement said that the programme of scientific and technical experiments to test the on-board systems and equipment had been fulfilled.

General Description An artist's drawing is available of Salyut at the time of writing; and descriptions of both interior and exterior have been issued, notably by Cosmonaut Dr Konstantin Feoktistov, who appears to have played a major part in its design. A docked Soyuz spacecraft appears to be an integral part of the complete Salyut system, with its orbital compartment providing sleeping accommodation; thus the somewhat generalized measurements appear to include the spacecraft. Salyut, says Feoktistov, looks like a cone with a docking unit at the end of it, followed by a small cylinder of about 6 ft 6¾ in. (2 m) diameter, expanding to 9 ft 10 in. (3 m) diameter, and a further widening to 13 ft 1½ in. (4 m). Finally there is a spherical bottom and a cone, containing the fuel tanks, with engine installations in a rear cylinder of roughly 6 ft 6¾ in. (2 m) diameter. Batteries are recharged by means of solar panels. Eight working positions are provided; in the 'passage-compartment', or docking tunnel, first entered by boarding cosmonauts, part of the scientific astrophysical apparatus and several control panels are provided. A hatch (at the 9 ft 10 in. (3 m) diameter expansion) leads from the tunnel to the main working compartment. Inside there is a small platform with two 'working seats' facing the hatch, with control and instrument panels in front of it, and Soyuz-type command and signal equipment at the side. Further back is a position for studying 'the plasma around the ship', and two more working positions concerned with the on-board equipment, regeneration units and filters; behind this is equipment for medico-biological research. The station was said to contain 'many tons of apparatus', including telescopes, spectrometers, electrophotometers and television devices. One of the main objects was to perfect the design of such instruments for use in future space stations. Orientation and guidance is carried out either by ground control or manually by the cosmonauts. A small lever on the central control panel, when turned right, activates small thrusters which rotate the craft to the right; when pulled back or pushed, the forward part, to which Soyuz is docked, is raised or lowered. An optical sight in one window enables the earth's surface or horizon, or objects in space, to be used as a guide for turning the solar panels towards the sun. Salyut, at any rate in its first version, contains only one docking port. The first, $5\frac{1}{2}$-hour docking with Salyut of Soyuz 10 on April 23, 1971, is fully described under **Soyuz** (q.v.). Salyut provides much greater personal comfort for the cosmonauts than previous spaceflights: there is a well-stocked refrigerator, with devices to heat both food and water; drinking water is stored in rubber containers, and kept potable for about 90 days with the addition of a small quantity of ionic silver. A Soviet commentator suggests that in future fresh water may be 'pumped' up to space stations in the form of steam passed along a laser beam.

SKYLAB—US MANNED EXPERIMENTAL SPACE STATION

General Cluster Data

	Length	**Dia**	**Weight**
CSM	34·3 ft	13 ft	30,800 lb
	(10·45 m)	(3·96 m)	(13,970 kg)
MDA	17·4 ft	10 ft	13,800 lb
	(5·30 m)	(3·05 m)	(6260 kg)
ATM	11·0 ft	7 ft	24,650 lb
	(3·35 m)	(2·13 m)	(11,180 kg)
AM (STS)	3·9 ft	10 ft	
	(1·19 m)	(3·05 m)	} (49,000 lb)
AM (TNL)	13·7 ft	5·5 ft	(22,225 kg)
	(4·18 m)	(1·68 m)	
IU	3·0 ft	21·6 ft	4,550 lb
	(0·91 m)	(6·58 m)	(2065 kg)
OWS	48·1 ft	21·6 ft	78,000 lb
	(14·66 m)	(6·58 m)	(35,380 kg)

Total Cluster Weight: 199,000* (90,265 kg)
Total Cluster Length: 118·5 ft (36·12 m)
Total Work Volume: 12,763 cu ft (361·4 m³)

* Excludes Payload Shroud: (26,000 lb; 11,795 kg)

General Description The Skylab Workshop is due to be launched on April 30, 1973, and to be manned by three crews, each of three astronauts; the crews will include doctors, engineers and scientists. Originally called the Post-Apollo Applications Programme, and intended to obtain the maximum return from left-over Apollo hardware, the project has gradually become a vital step in the development of man's exploitation of the space environment. The Mission will cover a period of 8 months; and the three crews will man the Workshop for three periods of up to 28, 56 and 56 days with intervals of about 1 to 2 months

Artist's drawing of Skylab

between flights. Nearly 26,000 scientists and other workers have been employed on preparations for the mission.

The objects are: 1) to extend man's capabilities to live and work in space, and to gain knowledge of the physiological and psychological effects on man of long periods of weightlessness; 2) to operate sophisticated scientific equipment to look outward at the sun to study solar activity; and 3) to look inward to earth to study man's most pressing environmental problems, such as pollution, flooding, erosion and the depletion of mineral resources.

The 270-mile (435-km) earth orbit, and 50° inclination in which the Workshop will be placed, will enable observations to cover the US, most of Europe and South America, and the whole of Africa, Australia and China. Over 2900 astronaut hours have been allocated to about fifty experiments, ten of which will be aimed at determining how the human body adjusts and performs under the conditions of prolonged spaceflights. So that physical changes due to weightlessness can be identified, the post-flight condition of the astronauts will be compared with that of three men who, it is planned, will spend 56 days in a simulated Skylab Workshop at Houston's Manned Spacecraft Centre, where it is possible to provide an identical low-pressure breathing atmosphere (70 per cent oxygen and 30 per cent nitrogen at 5 psi), plus a Skylab schedule of work, food and recreation, but not the condition of weightlessness.

The main Skylab Cluster will be launched by a Saturn 5 rocket, with the first crew being launched in an Apollo spacecraft by a Saturn 1B rocket one day later; the second and third crews will be launched approximately 90 and 180 days after the first launch, also by Saturn 1B rocket.

Though only one mission is so far budgeted, there is backup hardware in case of a launch failure. If finance were made available, this could be used for a second Skylab, which could be launched in 1974 to help fill the gap between these post-Apollo flights and the first manned orbital flight of the Space Shuttle, not now likely before 1978.

For the first time in a US manned flight, there will be a limited 'rescue capability', for use if one of the crews should find itself stranded in space with no ability to return. An Apollo Command Module is being prepared, able to carry two men into orbit and bring five back to earth—though the stranded astronauts would have to survive for up to 45 days, depending upon how far into the mission the need for rescue arose, but in any event, no longer than 10 days beyond the normal mission time.

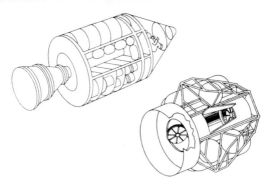

Skylab components *left:* Command and Service Module
right: Airlock Module

The Skylab Cluster consists of 6 main components:

CSM The Command and Service Module is basically the same vehicle used in the Apollo lunar missions. Minor modifications provide additional attitude control capability, and enable it to be made semi-dormant while docked to the cluster. The Environmental Control System, which provides an all-oxygen atmosphere at 5 psi (0·35 kg/cm²), is shut down when docked, and a ducted fan in the MDA hatch circulates Skylab's two-gas atmosphere. Operating power when it is docked is also received from Skylab's solar array power system instead of from the CSM's two fuel cells. There is special stowage provision for the return to earth of films, magnetic tape and biomedical specimens.

AM The Airlock Module provides a special hatch, airlock compartment and services for supporting EVA from the cluster when crew members retrieve and replace the ATM film. It serves as a passage-way between the OWS and MDA. The AM provides the main power distribution control and the atmosphere supply and conditioning equipment for the entire cluster. Electrical power comes from eight batteries charged from the OWS solar array.

MDA The Multiple Docking Adapter derives its name from the two ports which allow the Apollo spacecraft to dock to the Skylab. The end, or axial port, is for normal use; the side, or radial port, is for emergency use if the end port becomes unusable. Crews transferring from the CM to Skylab use the same mechanisms and procedures employed for transfer to the LM in the Apollo programme. The MDA is also the control centre and a working area for the ATM and Earth Resources Experiments; it provides the control and display consoles for both, and carries the remote sensors and cameras for the Earth Resources Experiments, vaults for storage of photographic film, and a work chamber for experimentation with materials processing in zero gravity. The Thruster Attitude Control System (TACS) is also housed in the MDA.

ATM The Apollo Telescope Mount is a solar observatory, placing large sophisticated telescopes beyond the obscuring atmosphere, enabling astronomer-astronauts to make detailed observations of the sun over long periods, and involving six specific experiments. A deployment mechanism swings it 90° from its launch position in the nose on top of the rocket so that it is at the side of the MDA, thus exposing the MDA's axial port for CSM docking. The four solar array wings, deployed at the same time, total 1200 sq ft ($111 \cdot 5$ m^2), providing $10 \cdot 5$ kW at $131°F$ ($55°C$) to recharge the

Plate 1 Launching Apollo 7, the first manned Apollo spaceflight

The first space walk, made by Ed White from Gemini 4

Plate 2

David Scott taking pictures of Russell Schweickart's space walk from the open Command Module hatch of Apollo 9

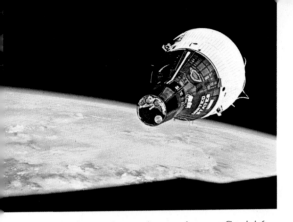

Plate 3 The first rendezvous in space between Gemini 6 and 7

Lunar Module in space during the Apollo 9 mission

Plate 4 Apollo 9 blast-off

Apollo 11's
Command
Module in
lunar orbit

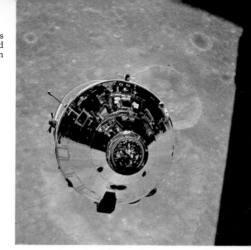

Plate 5

Apollo 11's
Lunar Mod-
ule descend-
ing on the
lunar sur-
face, photo-
graphed
from the
Command
Module
window

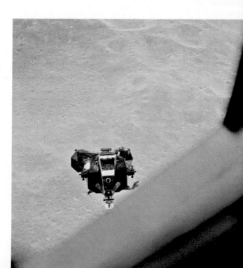

Plate 6 Edwin Aldrin, second man on the moon

James Irwin beside the US flag and the Lunar Rover in the Hadley Delta during the Apollo 15 mission

The Earth
photograph-
ed from space

Plate 7

Apollo 9's
Atlantic re-
covery

Plate 8   Vostok Rocket at the Paris Air Show

Soyuz 2 spacecraft docking, as shown at the Paris
Air Show

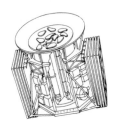

Skylab components *left:* Multiple Docking Adapter
right: Apollo Telescope Mount

eighteen ATM batteries from solar power.
Average output capability is approximately 3700
W on each orbit. The ATM contains six of Skylab's eleven telescopes. The aim is to increase our
knowledge of solar activity and its effect on the
earth's environment; to test the ability of man to
operate complex scientific instruments in a space
environment; and to gather engineering and
technical details for the development of such
systems. The ATM's large size accommodates
instruments 10 ft (3·05 m) long; a highly accurate
pointing system enables the astronauts to focus
quickly on any suddenly occurring solar activity.
The sun sensors, computers, and control moment
gyros (CMG) are essential parts of the attitude
control systems for the whole Skylab cluster.

OWS The Orbiting Workshop completes the
Saturn Workshop (SWS). Originally a third-
stage, S4B rocket, it has been converted into a two-
storey compartment providing 11,000 cu ft
(311·5 m³) of space. What was originally the
liquid oxygen tank at the bottom has been converted into a trash disposal system, or dustbin,
for storing waste matter. (A major problem for
orbiting space stations is to avoid pollution de-

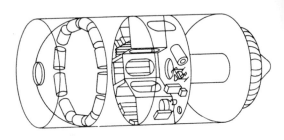

Skylab's Orbiting Workshop

veloping around the vehicle if waste matters are dumped outside.) The first floor, immediately above the trash disposal system, is used for crew quarters, with three compartments. The 'ward room' has a picture window which will be turned for observation purposes towards the earth. A central table provides facilities for preparing and heating food. There are food and storage racks, a refrigerator and deep freeze; there is a shower, in which the water is sucked down to overcome the weightlessness problem; and, for the first time in a spacecraft, a conventional toilet, on which astronauts must strap themselves as on an aircraft seat. There are handrails to assist movement and the astronauts wear special shoes (Foot Controlled Manoeuvring Units) with which, by a slight twist of the ankle, they can lock themselves into the grilled floors. The upper floor contains most of the experiments, and includes two scientific airlocks enabling experiments to be pushed out and exposed to space conditions. The pressure hull of the OWS is shielded against possible meteorite strikes. Solar array wings, 23 by 30 ft (7·01 by 9·14 m) on each side, provide 1200 sq ft (111·5 m²) of solar energy conversion area for charging the AM's eight batteries on each orbit. Six cold-gas thrusters are mounted in groups of three on the aft end for controlling changes in attitude.

IU The Instrument Unit. The Saturn 5 Instrument Unit is used only during launching and the first $7\frac{1}{2}$ hours of orbital operations. It provides tracking, command, measuring and telemetry systems and an electrical power supply used from lift-off to insertion into orbit. It then provides guidance and sequencing functions to deploy the ATM and OWS and their solar arrays.

Note The Payload Shroud provides an environmental shield and an aerodynamic fairing from lift-off to insertion. It consists of a spherical cap mounted on a double-angle nose cone. It has access openings with work platforms at four levels for pre-launch assembly. Shortly after the S2 stage has been separated following orbit insertion, explosive bolts are used to break the shroud into four leaf-like sections and blow them clear of the cluster.

Skylab Mission Profile

Flights 1 and 2

Day −1 The unmanned Saturn Workshop (SWS) will be launched by a two-stage Saturn 5 from Kennedy Space Centre's Launch Complex 39A. During the $7\frac{1}{2}$-hour lifetime of the Instrument Unit (IU) the protective Payload Shroud will be jettisoned, and the SWS will be automatically oriented to a solar-inertial, or sun-pointing, attitude. The Apollo Telescope Mount (ATM) will be rotated 90° from its folded launch position, and the four solar array panels, each 49 ft (14·94 m) long—in appearance rather like a four-bladed windmill—will be deployed into position to supply power to the whole cluster. The interior of the SWS will be pressurized to 5 psi (0·35 kg/cm²), with a mixture of 3·7 psi (0·26 kg/cm²) oxygen and 1·3 psi (0·09 kg/cm²) nitrogen—marking a final break by NASA with the all-oxygen environment used in previous manned flights. The SWS, containing all consumables for the whole mission, including oxygen, water, food and clothing, equivalent to 1 ton (1·02 tonne) for every man-month, will then be ready for occupation.

1 The first CSM will be launched by a two-stage Saturn 1B from Kennedy Space Centre's Launch Complex 39B, into an interim orbit of 93 by 183 miles (149·5 by 294·5 km). The CSM will use its 20,000 lb (9070 kg) thrust Service Propulsion System for boosting to the SWS orbit, and dock with the axial port of the Multiple Docking Adapter (MDA), thus completing the cluster. The crew will enter and complete activation of the SWS for habitation; the CSM will be powered-down to the maximum extent, with only essential elements of communications, instrumentation and thermal control systems operating.

2–25 Mission experiment operations. On this first mission, the primary objective will be a series of medical experiments, aimed at extending manned flights. (Previous longest US flight was Gemini 7, in 1965, lasting 330 hours 35 minutes). But ATM operation will be checked, and solar astronomy, earth-resources, and technical experiments started.

26 First EVA to retrieve ATM film, put in new film, and retrieve samples of experimental thermal coatings exposed to the space environment.

27–28 De-activate experiments, and close down SWS systems for non-operational, unmanned period of about 2 months.

28 CSM separation, deorbit, and return to earth, carrying about 1000 lb (454 kg) of experimental data, film, specimens and records. Splashdown in West Atlantic.

Flight 3

Day 1 Second crew launched, approximately 90 days after first launch. Rendezvous, docking and cluster activation identical to Flight 2.

2 – 3 Activate SWS experiments.

4 Second EVA to retrieve ATM film and reload.

Skylab Flight Missions

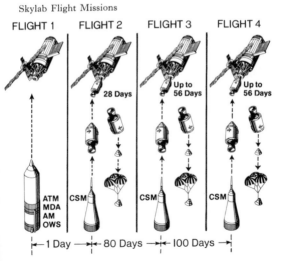

FLIGHT 1   FLIGHT 2   FLIGHT 3   FLIGHT 4

28 Days   Up to 56 Days   Up to 56 Days

ATM
MDA
AM
OWS   CSM   CSM   CSM

|← 1 Day →|← 80 Days →|← 100 Days →|

5–28 Mission Experiment Operations, with medical experiments aimed at evaluating effects on man of long-duration spaceflight being re-performed, but with more emphasis on solar astronomy, technical and earth resources experiments.

29 Third EVA for ATM film replacement.

30–53 Mission Experiment Operations.

54 Fourth EVA for film retrieval and repeat of Day 26, Flight 2, experiments.

55–56 Close down of SWS for second non-operational period of about $1\frac{1}{2}$ months.

56 CSM separation, deorbit and return to earth with film and specimens as before. Splashdown likely in mid-Pacific.

Flight 4

Day 1 Third crew launched approximately 90 days after start of Flight 3. Docking and Cluster Activation.

2–3 Start up Mission Experiment Operations, with main emphasis this time on earth resources activities.

4 Fifth EVA for ATM film replacement.

5–53 Mission Experiment Operations. Like earlier missions, it will be 'open-ended', and will be terminated earlier than 56 days in the event of illness or technical problems. But it is hoped the three missions together will provide valuable data on the reactions of nine different astronauts to long-duration spaceflight, and their ability to re-adapt after return.

54 Sixth EVA for ATM film retrieval and retrieval of thermal coatings.

55–56 Final deactivation of SWS systems, and preparation of Cluster for orbital storage. Orbital lifetime of SWS is about another 4 years.

56 CSM separation, deorbit and splashdown, probably in mid-Pacific, with final consignment of film and specimens.

Experiments Summary

Solar Astronomy Since the sun—our only nearby star—dominates the solar system, and is the single most influential element affecting life on earth, its study should enable us to learn more about the universe, space environment, and the solar system. The ATM experiments are the largest and most complex yet designed for an orbiting system; special emphasis is placed on emissions of the sun which are invisible on the ground because of absorption in the earth's atmosphere. The corona, the envelope of fiery gas surrounding the sun which is only normally visible during a total solar eclipse, will be monitored and photographed. X-ray telescopes will photograph X-ray emissions. The astronauts will visually scan the sun through onboard displays to locate areas of special interest, and will assist and advise ground-based astronomers in aligning and calibrating the ATM telescopes and instruments, putting them right, possibly by EVAs, if mechanical failures occur.

Earth Resources The Workshop will provide a platform in space from which to measure and observe a large part of the earth's surface. Six cameras will be used for multispectral photographic analysis of crop conditions, forestry, underground water, snow cover, water temperature and geological formations. Ground aircraft will conduct similar experiments at the same time for comparison. *Ocean surfaces* will be studied to establish global patterns of roughness, wave conditions and surface wind; areas of ice, rain and clouds will be identified, and sea surface temperatures deduced. The results should help in predicting both weather and ocean wave conditions. *Land surfaces* will be mapped to show snow cover, and its seasonable advances and retreats; changes in the borders of frozen and unfrozen ground; the seasonal changes of vegetation regions; and rainfall in remote regions.

Biomedical   These are planned to determine the effect of long-duration spaceflight upon the crews. Before and during the flight astronauts will be on a programmed diet so that changes in muscle and bone tissue—and particularly the calcium content of the latter—can be measured.   Changes in blood volume and red cells and the behaviour of the heart, urine and faeces input and output, will be studied during flight and after return to earth.   (There was considerable concern over the length of time it took the hearts of the Apollo 15 crew, following the first 12-day moonflight, to return to normal.) Studies of the astronauts' metabolic activities will provide information when planning ideal food supplies and working conditions for prolonged flights (such as would be involved in a Mars mission).   Mice and insects will be carried, and studies made of their body temperatures, heart rates and movements in weightless conditions.

Technology   These experiments are designed to make use of the space environment for the possible manufacture of materials in vacuum conditions and their retrieval for use on earth. Five experiments will study the feasibility of electron beam and thermo welding, molten metal flow, freezing patterns, thermal timing and surface tension for selected materials. The effects on materials of long-term space exposure will be studied; a laser-radar system will track the launch of the Saturn rockets to determine lift-off motion; and the gradual build-up of contamination around the Skylab cluster will be monitored to see if an atmosphere is gradually created around it, which could cause light-scattering and reduce the effectiveness of scientific observations.

Science An attempt will be made to analyse more than 1000 stars in the Milky Way by means of an Ultraviolet Panorama experiment funded and built by the French. The object is to study 'hot stars' distributed in different regions of the sky in relation to the Milky Way. The colour of each star must be established as a start to the process of classification. The skill and judgement of astronaut-observers will be severely tested by their study of twenty-five star fields, intended to obtain detailed physical analyses of individual stars and nebulae, and statistical studies of the frequency and galactic distribution of anomalous stars. Other experiments include a survey of a portion of sky to determine the source and intensity of X-rays, and the distances they have travelled; and photography of the airglow layer above the earth's atmosphere.

Skylab Crews

First Mission (May 1, 1973; 28 days) Commander, Charles Conrad, veteran of three flights, and commander of Apollo 12; science-pilot, Dr

Joseph Kerwin, a physician; and pilot Paul Weitz.

Second Mission (July 30, 1973; 56 days) Commander, Alan Bean, who accompanied Conrad on the moon on Apollo 12; science-pilot, Dr Owen Garriott, whose doctorate is in electrical engineering; and pilot, Jack Lousma.

Third Mission (October 28, 1973; 56 days) Commander, Gerald Carr; science-pilot, Dr Edward Gibson, whose doctorate is in engineering; and pilot, William Pogue. None of this crew has previous spaceflight experience.

Back-up Crews First Mission: Russell Schweickart, Dr Story Musgrave and Bruce McCandless. Second and Third Missions: Vance Brand, Dr William Lenoir and Dr Don Lind.

The dates given above are dependent, of course, upon Skylab going according to plan. A total of eighteen astronauts were assigned to Skylab, and the crews selected more than a year in advance to provide time for training in the scientific investigations as well as the mission. The scientist-astronauts, none of whom has previous spaceflight experience, were selected in June 1965, on the basis of their scientific qualifications, and two did not learn how to fly an aircraft until after becoming astronauts.

SOYUZ—USSR SPACECRAFT

Overall Length:	34 ft (10·36 m)
Diameter:	7 ft 6 in. to 9 ft 9 in. (2·29 m to 2·97 m)
Docking Probe:	9 ft (2·74 m)
Solar Wingspan:	33 ft (10·06 m)
Total Weight:	13,230 lb (6000 kg)
Total Volume:	318 cu ft (9 m³)

History Soyuz ('Union') is a long-term project aimed at the development of sustained flight, orbital manoeuvring, docking, scientific and technical investigations in earth orbit, and ultimately at the assembling of manned orbiting stations. It was designed by Chief Design Engineer Sergei Korolev, and his assistant, Voskresensky, shortly before their deaths (see page 132). The spherical and cylindrical shape of the Soyuz modules appear to be logical developments of their earlier vehicles. Soyuz is equipped for earth-orbit missions up to altitudes of about 800 miles (1300 km), and can be used either automatically or manually for both manoeuvring and docking (Plate 8). The system was successfully tested with an automatic docking of the unmanned Cosmos 186 and 188 satellites on October 30, 1967, which was the first Soviet docking of any sort to be achieved. This was successfully repeated with Cosmos 212 and 213 in April 1968. The series has encountered many setbacks as well as successes; the first trial flight of Soyuz 1, in April 1967, ended in disaster, (see page 111); the Soyuz 10 crew, after achieving Russia's first manned docking, were unable to enter the Salyut space station; and the Soyuz 11 crew, after a record 23 days in space, were killed by a catastrophic depressurization. Achievements, however, include the development of long-duration flight, the first space-welding experiments, and the first transfer of crews from one vehicle to another, in a simulated orbital rescue.

Spacecraft Description There are three compartments: a Re-entry, or Landing Module, in the centre, flanked on one side by an Orbital Compartment, used as a workshop and for resting while in orbit; and on the other side by an Instruments Compartment. The Orbital and Instrument Compartments are jettisoned just before re-entry.

Re-entry Module This is shaped like a car headlight, with an exterior coating as protection against re-entry heat, and interior insulation to provide both heat and sound protection. During re-entry interior temperatures do not exceed 25–30°C (77–86°F). The shape ensures aerodynamic lift, with deceleration normally producing no more than 3–4 G, compared with 8–10 G endured by cosmonauts during the ballistic re-entries of Vostok and Voskhod. The cosmonauts take their seats 2 hours before launch. The panel in front of the commander contains instruments and switches for the spacecraft systems, a TV screen, and optical orientation view-finder set up on a special porthole next to the panel. Orientation controls are on the commander's right, and manoeuvring controls on his left. There are portholes to starboard and port for visual and photographic observations. There are four TV cameras (two mounted externally and two inside), which provide 625-line transmission, twenty-five pictures per second. The atmosphere regeneration system contains alkali metals which absorb carbon dioxide and simultaneously release oxygen, maintaining a 14 lb psi (1 kg/cm²) conventional nitrogen/

Paris Air Show model showing the interior of a Soyuz spacecraft

oxygen atmosphere (compared with Apollo's all-oxygen atmosphere). Heat-exchange units condense excessive moisture and direct it to special moisture collectors. Water, as well as food, is carried in containers, since Soyuz does not use fuel-cells providing water as a by-product as in Apollo. Spacesuits are not considered necessary during launch and re-entry; comfortable leather jackets and helmets containing headsets are used, various types of spacesuits being donned during orbit for special purposes. Multi-channel telemetry systems store information in on-board memory units, and transmit it to earth during regular radio sessions. The hatch, in the upper part, or nose, is used for entry before launch and for transfer after launch to the orbital compartment. The re-entry procedure is started by a retrograde firing lasting about 146 seconds. The single main parachute, for which a back-up is available and which is preceded by a drogue, is deployed at 27,000 ft (8000 m); solid-fuelled retro-rockets, fired at about 3 ft (0·91 m) above the ground, ensure that landing velocity does not exceed 10 ft/sec (3 m/sec). A direction-finder transmitter sends out one signal during descent, and another after landing, to aid search parties if necessary.

Orbital Compartment Mounted on the nose of the completed craft, with an estimated volume of 221 cu ft (6·26 m³), this provides sufficient room for the cosmonauts to stand up, and an area for work, rest and sleep. It has four portholes for observation and filming, controls and communications systems, a portable TV camera, photo and cine-cameras. A 'sideboard' contains food, scientific equipment, medicine kit and washstand. Communications and rendezvous radar antennae are mounted on the exterior. It is used as an airlock by closing and sealing the hatch communicating with the Re-entry Module; when depressurized, the external hatch can be opened for egress into space. For entry into a Salyut space station, a hatch in the nose is used. In the case of an 'active' spacecraft, a 9-ft (2·74-m) docking probe is added to the nose; on a 'passive' craft, an adapter cone is fitted to receive the probe.

Service Module *Instrument Compartment* Cylindrical, about 9 ft 9 in. (2·97 m) diameter, this is at the rear, and like the Apollo Service Module, cannot be entered by the cosmonauts. It has a hermetically sealed instrument section housing the thermo-regulation system, electric supply system, orientation and movement control systems with a computer, and long-range radio communications and radio-telemetry. In the non-sealed section are two liquid-propelled rocket motors (the main and stand-by engines), each providing a thrust of 880 lb (400 kg). These are used for orbital manoeuvres up to a height of 800 miles (1300 km), and for braking purposes to start re-entry. A separate, low-thrust engine system provides attitude control. Mounted at the rear are two wing-like solar cell panels, each 12 ft (3·66 m) long; when deployed in orbit they provide a span of over 33 ft (10·06 m), and 150 sq ft (14 m²) of solar cell area to gather power for the spacecraft system. A single 'whip' antenna extends forward from the leading edge, near the tip of each panel.

Escape Tower For lift-off and launch, an escape tower is mounted on top of the Orbital Compartment, capable of lifting the Orbital and Re-entry Compartments clear of the rocket in the event of an emergency either on the launchpad or immediately after lift-off. It contains three separate tiers of rocket motors for boost, trajectory bending, and vernier control, to ensure that the spacecraft could be quickly removed from the trouble area.

Soyuz Flights

Soyuz 1 Russia's first manned flight for just over 2 years. Vladimir Komarov was launched on April 23, 1967, into a 139-mile (224-km) apogee, 125-mile (20-km) perigee orbit, with 51·7° inclination. Unofficial reports from Moscow had forecast that two spacecraft of a new type would be launched. The fact that for the first time a Soviet cosmonaut was being given a second flight, supported speculation that there might be attempts at docking and/or spacewalks. After the launch, the objects were said to be to check the systems and components of the new vehicle, to hold extended scientific and physical/technical experiments, and to continue medical and biological studies of the effects of spaceflight on the human organism. There appear to have been flight problems with Soyuz 1, however, and re-entry was ordered on the eighteenth orbit. At a height of 4 miles (6·5 km), the main parachute harness twisted, the spacecraft crashed to the ground, and Komarov was killed. He was the first man to have been killed during a mission in 6 years of space-flight. Earlier in the programme, centrifuge tests had shown that Komarov had developed an 'extra systole', or irregular heartbeat, and he had been suspended from the flight programme. Only after strenuous appeals had he been allowed to continue with command of Voskhod 1. There is no evidence, however, that he had any health problems during either of his flights. (Derek Slayton, one of the original seven Mercury astronauts, developed a similar heart condition; as a result, he was the only Mercury man never allowed to fly.)

Soyuz 2 and 3 Soyuz 2, used as an unmanned target vehicle, was launched on October 25, 1968, into a 139-mile (224-km) by 115-mile (185-km) orbit with 51° inclination; it was followed by Soyuz 3, manned by 47-year-old Giorgi Beregovoi, on October 26, into a 140-mile (225-km) by 127-mile (205-km) orbit. The Soyuz 2 launch was not announced until after Soyuz 3 was in orbit. Soyuz 2 was believed to be equipped with a docking collar, and Soyuz 3 made an automatic approach and rendezvous to within 600 ft (180 m), but there was no docking. The spacecraft then separated to 351 miles (565 km), and Beregovoi

The rocket carrier with spaceship Soyuz 9 in launch position

then carried out a second, manually controlled approach; no distance was given. On October 28 Soyuz 2 made an automatic re-entry, thus successfully testing the parachute system. Beregovoi continued his flight, making regular TV reports, his work including photography of the Earth's cloud and snow cover, and study of typhoons and cyclones. During orbit 36 an automatic manoeuvre changed the orbit to 151·5 miles (244 km) by 123·5 miles (199 km); then Beregovoi manually oriented the spacecraft, and switched on the automatic system for re-entry. The retro-rocket was fired for 145 seconds. The descent went so well that the search party was able to see and photograph the final descent, at the end of the sixty-four-orbit flight lasting 94 hours 51 minutes. With Soviet recovery ships stationed in the Indian Ocean, it was expected that Beregovoi would come down at sea, especially as Russia had made her first sea recovery of an unmanned spacecraft (Zond 5) in that area a month earlier. These preparations, however, were probably for use in emergency; after the successful landing of Soyuz 2 Beregovoi followed Russia's established landing procedures, and came down in a snowdrift in Kazakhstan. Asked afterwards whether his age had made it difficult for him to get accepted for this flight, Beregovoi said his height of 5 ft 11 in. (180 cm) had proved more of a problem than his age.

Soyuz 4 and 5 Russia's first manned docking, followed by the spacewalk-transfer of two cosmonauts from one craft to another, was achieved on this mission. Soyuz 4, piloted by Vladimir Shatalov, was launched on January 14, 1969 at an inclination of 51°, and its initial orbit of 140 miles (225 km) by 107·5 miles (173 km) was corrected on the fourth orbit to 147 miles (237 km) by 128½ miles (207 km).   This was in preparation for the launch of Soyuz 5 on January 15 into a 143-mile (230-km) by 124¼-mile (200-km) orbit; on board were three cosmonauts, Boris Volynov (Commander), Yevgeny Khrunov (Research Engineer), and Alexei Yeliseyev (Flight Engineer). On January 16, while Soyuz 4 was completing its thirty-fourth orbit and Soyuz 5 its eighteenth, the two vehicles were automatically brought within 328 ft (100 m); Shatalov then took over manual control, and docked Soyuz 4 with 5. Outside TV cameras transmitted the docking process to earth.   The two craft were coupled mechanically, and electrical connections and telephone communications established.   Volynov oriented the joined craft, which Russia claimed to be 'the world's first experimental space station', so that the solar arrays were exposed to the sun; four compartments, it was pointed out, had become available to provide comfortable conditions for work and rest for the combined crew of four. Immediately after the docking Khrunov and Yeliseyev donned a new type of spacesuit with self-supporting life systems, and egressed into space through the hatch of Soyuz 5's orbital module. Life-sustaining packs containing oxygen supplies and air-conditioning systems, were attached to their legs so as not to get in the way as they used external handrails during their 37-minute space-walk into the Soyuz 4 orbital compartment.

The two craft were undocked after 4 hours, and the following day Soyuz 4, on its forty-eighth orbit after 71 hours 14 minutes of flight, successfully re-entered and landed.   Helicopters sighted the orange parachute as the spacecraft touched down on target in the Karaganda area, in a strong wind and temperature of − 35°, and warm clothes were hurried to the site for the cosmonauts. Volynov, alone in Soyuz 5, successfully re-entered and landed on January 18, after fifty orbits and 72 hours 46 minutes. In effect, the mission had rehearsed the first emergency rescue in space. Leonov, the first spacewalker, said afterwards he had advised Khrunov and Yeliseyev, when spacewalking, 'to think ten times before moving a finger, and twenty times before moving a hand', since abrupt and hurried movements built up heat in the spacesuit and could make a cosmonaut unfit for work. This mission, following shortly after America's first flight around the moon with Apollo 8, saw a new style of Russian coverage. TV pictures, though still not released live, were shown only an hour after the launches, and were followed by TV tours around the spacecraft, conducted by Shatalov in Soyuz 4.

Soyuz 6, 7 and 8 These spacecraft were launched on successive days, and each remained in orbit for 5 days; the group flight, therefore, the first time three manned vehicles had been in orbit simultaneously, covered a total of 7 days. Soyuz 6, launched on October 11, 1969 into an initial orbit of 115 miles (185 km) by 138 miles (222 km), carried Georgi Shonin (Commander) and Valeri Kubasov (Flight Engineer). The inclination in the case of all three craft was 51·7°. Soyuz 7, launched on October 12, carried Anatoli Filipchenko (Commander), Viktor Gorbatko (Research Engineer) and Vladislav Volkov (Flight Engineer), into an initial orbit of 128 miles (206 km) by 140 miles (225 km). Soyuz 8, on October 13, carried Vladimir Shatalov (Overall Commander), and Alexei Yeliseyev (Flight Engineer), into an initial orbit of 127 miles (204·5 km) by 138 miles (222 km). Both crew members of Soyuz 8 were making their second flight.

Early in the mission it was stated that Soyuz 6 carried extra scientific, instead of docking equipment; the other vehicles were believed to be equipped for docking, but none took place, despite much speculation during the flight that 7 and 8 would dock, and spacewalks would follow. Some observers thought there had been manual control problems with the improved autonomous navigation systems on Soyuz 7 and 8. At the subsequent Moscow news conference, however, Filipchenko said no spacesuits had been taken. Shatalov said docking would have been possible, but was not included in the tasks; he did, however, say there had been 'difficulties, as in every spaceflight'.

Objectives given included 'mutual manoeuvring' to test the complex system of controlling group flight by three spaceships. For the first time Soviet ground tracking stations were sup-

Service towers on the Baikonur launchpad holding the rocket carrier and Soyuz 9

plemented by eight Academy of Science research ships on station in various parts of the world. A total of thirty-one orbital changes were made, mostly manually, by the crews, in a wide variety of rendezvous manoeuvres. On October 15 Soyuz 7 and 8 conducted a manual rendezvous to within 1600 ft (488 m) of one another, observed by Soyuz 5 from a distance of several miles. This position was maintained for about 24 hours. Continuous radio and radar contact was maintained among the spaceships, and with tracking ships and ground stations, through Molniya satellites. Crew tasks carried out included scientific and photographic studies of near-Earth space, and Earth photography for geological, geographic and meteorological purposes. The activities which attracted most attention, however, were experiments in remote-control welding; for the first time in their space programme, Soviet scientists announced in advance that they were to take place. When assembling space stations, it was explained, cosmonauts would have to work outside the spacecraft for long periods. Welding processes were much preferable to joining parts with nuts and bolts, since the latter required revolving movements, which were difficult for weightless cosmonauts to achieve. On their seventy-seventh orbit, Shonin and Kubasov, in Soyuz 6, sealed themselves in their re-entry module, and depressurized the orbital workshop, creating the conditions known to be ideal for welding. They used a remote control panel to operate the 'Vulkan' experimental equipment in the workshop. Weighing about 110 lb (50 kg), this consisted of a welding unit, turntable with specimens of welded metals, instrument unit with power pack, and safety shield to cover the welding unit. Kubasov tested three methods: compressed arc welding (low-temperature plasma); electron beam; and fusible (or consumable) electrode welding. The last two, it was stated afterwards, were the most promising methods because solar energy could be used to heat the components. After repressurizing the orbital compartment, Kubasov transferred samples of the experiments to the re-entry module, and Soyuz 6 landed after completing eighty orbits and 118 hours 42 minutes. Soyuz 7 landed one day later, on October 17, after eighty orbits and 118 hours 41 minutes; and Soyuz 8 on October 18 after eighty orbits and 118 hours 41 minutes.

Soyuz 9   This established a new long-duration record of nearly 18 days' (424 hours) spaceflight, surpassing the previous record established 5 years earlier by Gemini 7. It was the first manned launch made at night. The two-man crew was Andrian Nikolayev, Commander, who had to wait 8 years for his second flight; and Vitali Sevastyanov, a civilian, as Flight Engineer. Launch on June 1, 1970, was into a 138- by 129-mile (222- by 207·5-km) orbit, with 51·7° inclination, increased by manual firings on the fifth and seventeenth orbits to 166 by 154 miles (267 by 248 km). The primary aim was to study the physical effects of prolonged weightlessness; other work, however, included the study and photography of ocean behaviour, including coastal currents and surface temperatures, aimed at developing fish-searching techniques employed by Soviet fishing fleets. During the flight cabin pressure was maintained at an average of 16 psi (1·12 kg/cm^2), or slightly higher than sea level, with 23·4% oxygen and the remainder nitrogen and other gases; average temperature was 71°F (22°C). Final descent was shown on Soviet TV—the first time the final stages of a Russian recovery had been seen in this way. Soviet doctors' concern about long-term effects appeared justified by the fact that both cosmonauts had difficulty in re-adjusting to earth conditions. For 5 to 8 days they said they felt as if they were in a centrifuge, being subjected to 2 G (that is, feeling as if they were twice their actual weight); they found difficulty in walking, and sleep in bed was uncomfortable for 4 or 5 days. During the flight Nikolayev's daughter, Elena, was brought to Ground Control to talk to him to celebrate her sixth birthday; she was born a year after Nikolayev married Valentina Tereshkova, the first woman in space, following their original flights.

Soyuz 10 Launched 3 days after Salyut 1 (q.v.), Russia's first experimental space station, Soyuz 10 had a three-man crew: Vladimir Shatalov (Commander), and Alexei Yeliseyev (Flight Engineer), both making their third flights; and Nikolai Rukavishnikov (Test Engineer). It was a pre-dawn launch on April 23, 1971 into a 154- by 130-mile (248- by 209-km) orbit, and inclination of 51·6°. Mid-course corrections, transmitted out of range of Soviet territory by the research ship *Sergei Korolev*, started the automatic rendezvous procedure. Yeliseyev said the cosmonauts first saw Salyut on an 'optical instrument' at a range of 9 miles (15 km). When the automatic procedures had brought Soyuz 10 to within 600 ft (180 m) of Salyut, the crew took over manual control, and Shatalov completed the final approach and docking during the twelfth orbit of Soyuz 10 and the eighty-sixth orbit of Salyut. Cosmonaut Dr Boris Yegorov (who flew on Voskhod 1) observed later that the manoeuvre 'led to a considerable emotional load on the cosmonauts'.

Soyuz 10 on its way to the launchpad

The two vehicles remained docked for $5\frac{1}{2}$ hours, and official statements said the principles of closing and berthing the craft, the operation of the new coupling units, and the complex radio engineering equipment were tested. The crew did not enter Salyut, and Western speculation was that either they had been unable to open the docking tunnel, or that the flight had been cut short because Rukavishnikov became ill. (Rukavishnikov had said during a communications session that the presence and advice of Shatalov and Yeliseyev 'helped him to get accustomed to weightlessness, and to overcome the unusual and rather unpleasant sensations arising as a result of the increased flow of blood to the head'.) Soviet reports however, emphasized that Shatalov had carried out more than ten complex manoeuvres during rendezvous, berthing, docking and undocking, and that docking had had to be done independently while the crew was 'out of radio contact with ground stations on Soviet territory'. (Ships were apparently not mentioned on this occasion.) It was also pointed out that docking a small spacecraft with a vehicle of much greater mass was more difficult than docking two Soyuz or Cosmos craft. Soyuz 10 remained in orbit for 16 hours after undocking; then, after thirty orbits, re-entered and landed in darkness (2.40 a.m. local time) 75 miles (120 km) north-west of Karaganda on April 25. At a subsequent news conference Rukavishnikov, asked about Western reports of his health, said 'work in a weightless state was a joy'; after the second day he had been too busy even to notice any unpleasant sensations. Reasons for the cosmonauts not entering Salyut, which continued its flight after Soyuz 10 re-entered, were apparently not discussed.

Soyuz 11 The three-man crew made history by spending nearly 24 days in space—23 of them working in Salyut 1—but they died during re-entry owing to the failure of a hatch-seal on the spacecraft. Soyuz 11, crewed by Georgi Dobrovolsky (Commander), Vladislav Volkov (Flight Engineer), making his second flight, and Victor Patsayev (Test Engineer), was launched on June 6, 1971, and following a fourth orbit correction, given an apogee of 135 miles (217 km) and perigee of 115 miles (185 km), and 51·6° inclination. The purpose of the mission was 'to continue comprehensive scientific and technical studies in joint flight with the Salyut orbital scientific station'. On entering orbit, Soyuz 11 was 1864 miles (3000 km) behind Salyut; the following day, on the sixteenth orbit, a second manoeuvre initiated rendezvous, and automatic approach was carried out from 3·7 miles (6 km) to 328 ft (100 m). The crew then took over, completing the docking manually, with the two craft in a 135- by 115-mile (217- by 185-km) orbit. The total weight of the two craft was over 25 tons (25·4 tonnes). The vehicles were rigidly coupled, and electrical and hydraulic communications connected. The hatches of the airtight passage were opened and the cosmonauts entered the scientific station. It was announced that Soyuz spacecraft had been assigned the role of ferry-craft for orbital stations, with the task of carrying relief crews and supplies, and returning the results of experiments to earth.

The preparation of Salyut as a manned orbital station took two days, during which the cosmonauts complained about having too much work. Soyuz 11 was then powered down for 23 days while work continued aboard Salyut (q.v.). On June 29 after, it was stated, completing their flight programme in full, the cosmonauts transferred their research materials and logs to Soyuz 11, and prepared to return to earth. Undocking took place at 9.28 p.m., the crew reporting that separation was normal, with all systems functioning. At 1.35 a.m. on June 30 (a record 23 days 17 hours

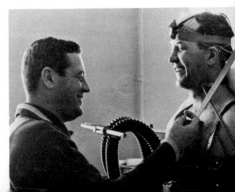

Soyuz 11's commander Dobrovolski undergoing a medical test; left is Volkov, flight Engineer

40 minutes, and 380 orbits since launch), a normal firing of the braking engine, usually 146 seconds, was carried out. From that moment, all communications with the crew ceased; but the automatic re-entry procedure of aerodynamic braking, parachute deployment, and soft-landing engines continued smoothly, with an on-target touchdown. A helicopter-recovery crew, landing simultaneously with the spacecraft, opened the hatch and found the crew lifeless in their seats. Although telemetry had shown the cosmonauts to be well after their long period in space, they had reported feeling fatigued on the day of their return; consequently many newspapers carried banner headlines speculating on the possibility that the debilitating effects of their long period of weightlessness had led to the cosmonauts being unable to withstand the sudden return to G-forces during re-entry. A Soviet Government Commission, however, announced 12 days later that a rapid depressurization in the re-entry vehicle, as a result of a loss of sealing, had led to the cosmonauts' deaths. They added that an inspection of the vehicle showed there were no failures in its design. It seems, therefore, that for some reason the hatch between Soyuz 11's orbital and re-entry modules was either faulty or had not been correctly sealed, so that when the orbital and systems compartments were jettisoned just before re-entry began, the air in the re-entry compartment rushed out, creating a catastrophic decompression. On July 2 the cremated remains of Lt-Col Dobrovolsky, aged 43, of Victor Patsayev, who celebrated his thirty-eighth birthday aboard Salyut, and Vladislav Volkov, aged 36, were placed in the Kremlin Wall, following a Moscow funeral attended by Brezhnev, Kosygin and other Soviet leaders.

SPACE SHUTTLE—US
SPACECRAFT

Lift-Off Thrust: 5·5 million lb (2·49 million kg)
Lift-Off Weight: 4·7 million lb (2·13 million kg)
Orbiter: Length: 120 ft (36·6 m)
          Wingspan: 75 ft (22·9 m)
          Payload: 65,000 lb (29,500 kg)
          Payload Diameter: 14–15 ft (4·3–4·6 m)
          Payload Length: 45–60 ft (13·7–18·3 m)
Booster: Length: 175 ft (53·3 m)
Development Cost: £2,291,600,000 ($5500 million)
Orbiter Cost: £104,125,000 ($250 million)
                each

History As the Apollo programme steadily established the practicability of space exploration NASA studied methods of creating a Space Transportation System (STS) which, in the words of President Nixon, would 'transform the space frontier of the 1970s into familiar territory, easily

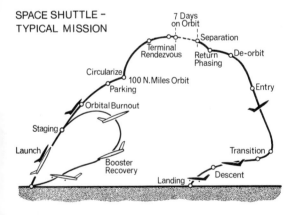

SPACE SHUTTLE –
TYPICAL MISSION

accessible for human endeavour in the 1980s and 1990s.' The original aim was to create a spacecraft which could be used over 100 times, and a manned booster which could also be flown back to base for re-use. Financial problems proved greater than technical ones: although the technicians would sooner press on with a system in which both spacecraft and booster were fully recoverable, since this would be more efficient in the long-term, it was decided in early 1972 to develop an aircraft-like orbiter, about the size of a DC9, which would be launched vertically like a rocket, fly in orbit like a spaceship, and land like an aircraft; and an unmanned 'semi-recoverable' booster the length of a Tristar which would descend by parachute and be recovered from the sea after use. A 6-year development plan provides for first manned flights in 1978, with the system becoming operational shortly after. Programme management has been given to Manned Spacecraft Centre (MSC), Houston, which is also responsible for the orbiter stage; Marshall Space Flight Centre, Huntsville, is responsible for the booster stage and shuttle engine; and Kennedy Space Centre for launch and recovery facilities. The estimated £2,291,600,000 ($5500 million) costs (about one quarter of the Apollo programme) include research, development, and two flight test vehicles. Each additional orbiter will cost £104,125,000 ($250 million); each additional booster £20,800,000 ($50 million). Another £125,000,000 ($300 million) will be spent on facilities. The cost of each flight is estimated at £4,160,000 ($10 million). Apollo 15 cost £185,416,000 ($445 million); and the hope is to reduce the cost per pound of putting a payload into space from £250–290 ($600–700) at present to £40 ($100). (Explorer 1, weighing 30 lb (13·6 kg),

Artist's impression of a Space Shuttle launch. Note Booster's external fuel tanks

cost £41,600 ($100,000) per pound.)

General Description At launch, the orbiter will be joined to the booster in piggy-back fashion. Separation altitude will be about 40 miles (64·5 km) and about 8000 mph (12,875 kph); the orbiter engines will then be ignited to take the speed to about 18,000 mph (28,968 kph) and place it in an orbit 100–700 miles (161–1125 km) high. Its lightweight aircraft structure must repeatedly withstand temperatures up to 3000°F (1650°C) during re-entries. It can land on runways about the length used by conventional commercial jets, with comparable postflight inspection, refurbishment and refuelling. Turnaround time to next launch will be about 2 weeks.

Orbiter Flown by a two-man crew, it will be able to remain in orbit from 1 week to 30 days, depending on mission requirements. Normally there will be provision for two passengers; but six to twelve, or even more additional passengers can be carried in special modules fitted into the payload bay. There are three high-pressure, liquid-oxygen, liquid-hydrogen engines, each developing 415,000 lb (188,240 kg) thrust and with propellants carried in an external tank about 125 ft (38 m) long, which will be jettisoned in orbit. About thirty-three attitude control thrusters will be in pods on the wingtips and fin. Early versions of the orbiter are also likely to be fitted with a retractable turbojet engine for landing. Since this would severely limit the size and weight of sortie modules, it is not planned to carry a landing engine on eastward launch missions. In these cases, 'deadstick landings' would be required, necessitating altitude corrections, flare and automatic landings; equipment for this may be based partly on the ground and partly in the spacecraft. Design aims at providing a multi-purpose vehicle, able to replace nearly all existing expendable launch vehicles; it will be used to carry most US payloads into space—both manned and unmanned, civilian and military; it will also provide for future needs of commercial users, and for both US and foreign government requirements. As a ferry, it will carry passengers and freight between earth and orbiting space laboratories which will be largely built from modules delivered by the shuttle system. Among the first missions will be deployment of weather and communications satellites, and repair and retrieval of automated spacecraft already in orbit. If necessary the shuttle could be available for rescue missions within 24 hours.

The need for this flexibility results in the large payload compartment, possibly up to 15 by 60 ft (4·50 by 18·25 m). Hatches on top of the compartment will open wide in orbit to facilitate un-loading of cargoes and deployment of modules or large spacecraft carried inside. The interior will be pressurized, so that passengers and crew will not need spacesuits. Acceleration limits of 3 G during launch and re-entry, and only brief indoc-trination periods (2 to 3 months) will mean that scientists, doctors, artists and photographers etc.— both men and women—can commute to and from space stations and laboratories.

Booster Overall length of about 175 ft (53·4 m) is about half the height of a Saturn 5. In March 1972 it was finally decided to use solid-rocket motors firing in parallel with (at the same time as) the orbiter motors at lift-off. The solid-fuel sys-tem was selected rather than a pressure-fed, liquid-fuelled booster, on the grounds that development costs would be reduced by £132,575,700 ($350

Artist's concept of the Space Shuttle Orbiter launching a satellite

million) and the technical risks would also be lessened. The aim is to develop a 156-in. (3·96 m) rocket, although the largest solid rocket flown in space so far is 120 in. (3·05 m) in diameter. Simultaneously with these decisions it was announced that the initial launch and landing site would be at the Kennedy Space Centre. Towards the end of the decade, a second launch and landing site would be phased in at Vandenberg, California, for flights requiring high inclination orbits.

The recoverable booster will be structurally designed to withstand an impact of 102 mph (164 kph) in a nose-down entry into water. Structure stiffening will be assisted by 300-psi (21 kg/cm²) pressurization, using for the purpose the inert gases needed during launch to force liquid-hydrogen propellant from the tank. A parachute system will provide the nose-down entry, with the booster nose designed to ensure clean entry, so that following impact the booster does not pop out of the water and fall on its side, causing structural damage.

Conclusion A launch rate of 100 per year would require five orbiters. In addition to direct savings resulting from the re-usable Shuttle system, NASA expects to achieve significant economies following the reduction in the number and types of launch vehicles needed for the national space programme. As the cost comes down, the use of space for commercial and government reasons will rapidly increase, and new uses will be discovered. Activities are already expected to embrace extended weather predictions, improved communications and manufacturing processes. Studies of the solar system could lead to the harnessing of the sun's energy as a source of pollution-free energy.

VOSKHOD—USSR SPACECRAFT

Weight—Voskhod 1: 11,728 lb (5320 kg)
         Voskhod 2: 12,527 lb (5682 kg)
Size: Not given.

History   Controversy as to the origin and purpose of these two flights has never been completely resolved. One view is that the first resulted from pressure brought by the then Soviet Prime Minister, Mr Krushchev, on the Chief Designer, Sergei Korolev, to perform a three-man flight before America flew a two-man Gemini spacecraft. This would certainly explain why, although Voskhod is officially described as 'different from Vostok in both structure and equipment', no pictures have ever been released; it would also line up with suggestions that it was so difficult to cram three cosmonauts into a stripped-down Vostok that it was necessary to fly them without spacesuits. However, these arguments cannot detract from the value and courage of Leonov's first spacewalk during Voskhod 2. The international sensation it caused—coupled no doubt with Leonov's demonstration of its relative safety and feasibility—led to a NASA meeting at Houston 11 days later (March 29, 1965). There it was decided that on Gemini 4, the following June, preparations should be made for Edward White to carry out a similar spacewalk, although up to that stage the plan had merely been for the spacecraft hatch to be opened so that White could stand up, with head and shoulders protruding into space, without actually leaving the vehicle. The official Soviet description of the objects of the Voskhod missions were 'to test the new multi-seater spaceship, to investigate the work capacity of a group of spacemen specialized in different spheres of science and engineering, to make physical and technical experiments, and to perform an extensive medico-biological investigation programme.'

The Voskhod flights proved to be the last made under the leadership of Sergei Korolev, Russia's Chief Design Engineer for rockets and spacecraft throughout the development of orbital flight. He died on January 15, 1966, 10 months after Voskhod 2, aged 60; only after his death was his identity publicly revealed. Korolev's health, like that of his deputy, Voskresenky, who died aged 52 after the Voskhod 1 mission, had been undermined by 6 years of imprisonment under the Stalin regime. Posthumous recognition came with the public burial of his ashes in the Kremlin wall. Cosmonauts present included Komarov, whose ashes joined those of Korolev less than a year later as a result of the Soyuz 1 accident.

Spacecraft Description   No exterior pictures of Voskhod ('Sunrise') have ever been issued; all the evidence, including the weights, suggests that they are basically the Vostok spacecraft modified for multi-man flights. The main change is the addition of a soft-landing system, providing retro-rockets to assist the parachutes just before touchdown. This makes it possible for the cosmonauts to remain in the spacecraft for the landing, instead of ejecting and making their final descent on individual parachutes.

Voskhod 2, although it carried only two men instead of three, was 1235 lb (560 kg) heavier than Voskhod 1; this would no doubt be due to the fact that this time spacesuits were essential, and in addition, an airlock had been fitted. This was tube-shaped, about 6 ft (1·83 m) long and 3 ft (0·91 m) in diameter, presumably fitted to the inside of the exit hatch, to form a chamber which the cosmonaut, having crawled inside, could depressurize before opening the external hatch. The interior was said to have two comfortable armchairs, upholstered in white, with two instrument panels overhead; one of them was for the airlock chamber.

The spacecraft's control panel includes 'a long handle to operate the manual orientation system'. A red metal hood covers a small black button marked 'Descent TDU'—the Russian initials for the retro-engines. To the left of the cabin, an instrument board includes a revolving globe, which continually indicates the spacecraft's exact location. The lenses of the cine and TV cameras are trained down from above.

Voskhod Flights

Voskhod 1 The first three-man flight, and the first and only flight, at the time of writing, on which a medical man has flown. Vladimir Komarov (pilot), Konstantin Feoktistov (scientist), and Boris Yegorov (physician), were launched on October 12, 1964, and completed sixteen orbits in 24 hours 17 minutes. The apogee was 254 miles (409 km), perigee 110·5 miles (178 km), and the inclination 65°. The higher orbit was possible because a reserve retro-rocket had been added; Vostok orbits were planned so that atmospheric resistance would ensure re-entry within 10 days in the event of retro-rocket failure. For the first time, the cosmonauts had no spacesuits; they wore 'light grey sports suits and special white space helmets'. The cosmonauts emphasized that they were much more comfortable; but the absence of spacesuits was almost certainly due to the need to save weight and space when fitting three men into the spacecraft. Dr Yegorov's seat was apparently set above and in front of the other two.   **Despite the** cramped conditions, however, Dr Yegorov, using detachable biosensors, was able to study the effects of the flight on his companions, while Feoktistov carried out some experiments on the behaviour of liquids in weightlessness, in addition to geophysical and astronomical observations. The usual telephone conversation between the spacecraft and the Soviet Prime Minister was notable because it proved to be Mr Krushchev's last public statement; while talking to them he observed that Mikoyan was 'pulling the receiver out of my hand'.   Mr Krushchev was displaced the day after the landing, and the usual 'Hero's Welcome' in Red Square was delayed so that his successors, Mr Brezhnev and Mr Kosygin, could attend.

Voskhod 2   Pavel Belyayev and Alexei Leonov were launched on March 18, 1965 into the highest apogee yet obtained in manned flight; apogee was 308 miles (495 km), perigee 107 miles (172 km), inclination the usual 65°.   The flight lasted 26 hours 2 minutes, and seventeen orbits.   Leonov later described how, over the USSR on the second orbit, he donned his spacesuit, with the help of Belyayev, entered the airlock and inflated his suit. The pressure he gave later as 0·4 atmospheres, or about 6 psi (0·42 kg/cm^2).   When he emerged he pushed himself away from the spacecraft, and he could clearly distinguish the Black Sea with its very black water and the Caucasian coastline.   He began rotating on his 16-ft (4·88-m) tether ten times a second, but did not lose orientation, and was able to maintain communications both with earth and spacecraft.   His pulse rate was 150–160 when he left the spacecraft, and peaked to 168.   When he pulled too vigorously on the tether

Leonov making man's first spacewalk from Voskhod 2

he had to put out his hands to avoid collision with the spacecraft. After 10 minutes, during which his manoeuvres were watched by TV, Leonov was instructed to return to the airlock; unofficial reports say he then ran into difficulties for the first time. His spacesuit had 'ballooned', an effect predicted by British aerospace scientists, with the result that it took 8 minutes of struggling before he was able to force his way in. Later in the flight the cosmonauts reported sighting an unidentified earth satellite.

A major crisis occurred on the sixteenth orbit, when the automatic re-entry system failed. For the first time it became necessary for a Soviet spacecraft to make a manual re-entry; an extra orbit was flown, while Belyayev made preparations to fire the retro-rocket himself; when he did so, he was at first uncertain about the spacecraft's attitude, but it proved to be correct. The landing was made in deep snow in a forest near Perm, about 1200 miles (2000 km) north of the planned area; the cosmonauts had to wait $2\frac{1}{2}$ hours for the first helicopter, and then had to stay overnight at Perm before being flown back to base. Leonov suffered no ill effects from his spacewalk; and so far as is known, neither did Belyayev, but he died as a result of internal trouble, on January 11, 1970, aged 37.

VOSTOK—USSR SPACECRAFT

Orbit Payload (Spacecraft and
  Last Stage): 13,602 lb (6170 kg)
Spacecraft Weight: 10,400 lb (4730 kg)
Re-entry Vehicle: 5291 lb (2400 kg)
Spacecraft Diameter: 7 ft 6½ in. (2·3 m)
Length (Spacecraft and Instrument
  Cylinder): 24 ft 1¼ in. (7·35 m)

History Manned spaceflight began with Yuri
Gagarin's one-orbit flight in Vostok 1 on April 12,
1961. The name Vostok ('East') applies both to
the spacecraft and the launch vehicle, but the
latter is dealt with separately under 'Launchers'.
Although the spacecraft were designed for automatic operation throughout the series, the cosmonauts' flight programme involved numerous
astronomical and geophysical studies, in addition to
the development of manned spaceflight techniques.
Observations of constellations, photographs of the
sun and the disc of the earth both at daybreak and
sunset, were included. On Vostoks 3–6 seeds of
higher plants, bacteria and human cancer cells
were carried, for later study of the effects of spaceflight.

Soviet preparations for the first manned flight
really started with Sputnik 2 in November 1957,
when the dog Laika was placed in orbit and his
behaviour monitored for 7 days. Recovery was
not attempted, and he died in space; Sputnik 4,
launched on May 15, 1960, saw the launch of the
first, unmanned Vostok prototype; the recovery
attempt failed, and the re-entry section remained
in space until its orbit decayed and it was burnt up
on re-entering 5 years later. Sputnik 5, the
second Vostok trial, in August 1960, was successful, and two dogs, Belka and Strelka, were

ejected and recovered in a parachute-borne container after eighteen orbits; the third Vostok trial (Sputnik 6), in December 1960, was successfully orbited, but again the recovery system failed. The trials were repeated with two more Vostoks (Sputniks 9 and 10) in March 1961, and this time dog passengers were successfully recovered on both occasions. Eighteen days after the second of these dogs, Zvezdochka, had flown, Russia's first man went into space.

Spacecraft Description Vostok consists of a relatively simple re-entry sphere, heat-shielded all round, attached to a cylindrical instrument section and re-entry rocket engine. Designed for operations up to 10 days, it has manual as well as automatic control, though basically intended to be controlled from the ground. The retro-rocket engine is used to reduce the orbital speed and start the vehicle on a descent trajectory. Environmental control maintains an oxygen-nitrogen atmosphere at sea-level pressure.

Soviet concern about the possible effects of weightlessness from the start of spaceflight was demonstrated by the fact that in Vostok 3, cosmonaut reactions were monitored by two television cameras, one transmitting a full-face view, and the other a face view in profile. Also starting with Vostok 3, electro-encephalogram, electro-oculogram and galvanic skin reactions were monitored throughout the flights.

Vostok Flights

Vostok 1 Gagarin was launched from Tyuratum (though official Soviet records still give its starting point as Baikonur, 230 miles (370 km) to the south-east) at 09.07 Moscow time on April 12, 1961,

Yuri Gagarin in the cabin of Vostok 1 before the flight

and landed 1 hour 48 minutes later at the village of Smelovka, Saratskaya. His orbit had a perigee of 112·5 miles (181 km), and apogee of 203 miles (327 km). Gagarin appears to have remained inside the spherical capsule for the landing, although the five subsequent Vostok cosmonauts all used ejection seats at 23,000 ft (7000 m), presumably because of the danger of the spacecraft bumping too heavily when it hit the ground. It was 4 years before pictures of Vostok were released, and the secrecy surrounding the early Soviet flights, never fully relaxed, makes it difficult even today to establish with certainty precise details of the flight.

The handsome Gagarin, with an engaging personality, ready smile and quick sense of humour, had clearly been chosen for the first-class public relations job he carried out following the flight. His first Western visit was to England, where he lunched with the Queen at a time when East-West

A mock-up of the Vostok spaceship exhibited in Moscow

relations were particularly strained, and went on to make appearances in many other countries. He was only 34 when he was killed, with another pilot, in what seems to have been a routine training flight in a jet aircraft.

Vostok 2 The first one-day flight. Herman Titov, launched on August 6, 1961, completed seventeen orbits in a flight lasting 25 hours 18 minutes. Apogee were 151·5 miles (244 km), perigee 113·5 miles (183 km), and inclination 65°; the orbit and inclination were similar for all the Vostok flights. A major factor in the Soviet decision to go straight from one orbit to seventeen was the problem of a suitable landing site on Soviet territory if fewer orbits were done. Spacecraft improvements were said to include a more advanced air-conditioning system. Titov, who was

26, emerged with credit from man's first full day in space, and from the subsequent news conferences and hero's parade in Red Square. Only later was it disclosed that he had suffered serious disorientation during the flight, and had inner ear trouble for some time afterwards.

Vostoks 3 and 4   The first 'group', or double flight, followed one year later. Andrian Nikolayev, launched in Vostok 3 on August 11, 1962, completed sixty-four orbits in 94 hours 27 minutes; for the last 3 days he was accompanied in orbit by Pavel Popovich, launched on August 12 in Vostok 4, who completed forty-eight orbits in 70 hours 29 minutes. Popovich, although he became the fourth Soviet cosmonaut to fly, had in fact been the first to be appointed. At one time the two spacecraft were within 3·1 miles (5 km) of one another. This has been attributed to the reliability of the Vostok launcher rather than to the sophistication of the spacecraft. Though at the time Western speculation was that the flight indicated that Russia was already moving towards rendezvous and docking techniques, there is no evidence that the Vostok spacecraft had any independent manoeuvring capability. The flight was notable for the first TV transmissions from space, and the first demonstrations of weightlessness seen by the public. Popovich was later said to have suffered some disorientation. The two cosmonauts, having used their ejection seats following simultaneous re-entries, finally landed only 120 miles (193 km) apart within 6 minutes.

Vostoks 5 and 6   The second group flight. Valeri Bykovsky, launched on June 14, 1963 in Vostok 5, established a space record of nearly 5 days (119 hours 6 minutes) and eighty-one orbits which stood for more than 2 years until broken by Gemini 5. World attention, however, was focused on Vostok 6; launched 2 days later, this carried the world's first spacewoman, Valentina Tereshkova, aged 26, who completed forty-eight orbits in 70 hours 50 minutes. According to some reports, Tereshkova was substituted at the last moment for a much more highly trained woman pilot who became indisposed. It is known that she suffered some disorientation and space sickness, but regular TV pictures relayed to earth, and exchanges in which she took part with Bykovsky and the ground did not fully accord with one report (by Vladimir Leonid) that 'when she landed in the Southern Urals she was in the most pitiful condition'. Tereshkova had limited space experience, and said herself soon after the flight that at the time of Titov's flight she was 'not even dreaming of becoming a cosmonaut'; she had, however, made 126 parachute jumps before Vostok 6.

Five months after her flight she married Nikolayev in Moscow, with the Soviet Prime Minister, Mr Kruschev, leading the wedding festivities. The healthy daughter born to the two cosmonauts within a year was an added bonus for Soviet scientists, still much concerned with possible radiation damage as a result of spaceflight. Apart from the considerable prestige associated with the first spaceflight achieved by a woman, this mission appears to have differed little technically from that of Vostok 4 and 5; however, the spacecraft were reported to have improved control systems, and further experience was gained in orbital rendezvous, and the effects of extended periods of weightlessness.

AGENA—US TARGET VEHICLE

Atlas-Agena Height:	104 ft (31·70 m)
Agena/Launch Height:	37 ft (11·28 m)
Agena Orbit Length:	26 ft (7·92 m)
Engines—Main	16,000 lb (7260 kg) thrust
SPS:	$\begin{cases} 2 \times 200 \text{ lb } (90·5 \text{ kg}) \text{ thrust} \\ 2 \times 16 \text{ lb } (7·25 \text{ kg}) \text{ thrust} \end{cases}$

History Agena upper stage rockets have been used for both classified and unclassified missions since 1959. The Agena-D version, launched as the 2nd stage of an Atlas, was used as a docking target on five of the Gemini flights. The first, on

Agena target seen from Gemini 8 during the latter's approach for the rendezvous

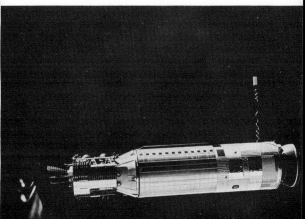

Gemini 6, was lost during launch; on the third launch, for Gemini 9, docking was impossible because the protective shroud around the docking ring failed to open properly, giving the Agena the famous 'angry alligator' look. On Gemini 8, however, history's first space-docking was achieved with the Agena target by Armstrong and Scott. Agena was selected as target vehicle because of its record of achieving precise predetermined orbits, and the fact that, once in orbit, it could be successfully stabilized and then maintain constant orientation to the earth. On Agena 8 it became the first spacecraft to be commanded and controlled by another space vehicle; the astronauts could control it, whether docked or merely in its immediate vicinity. Its main, gimballing engine, was restarted eight times by ground command after Gemini 8 had returned to earth. Gemini 10 saw the first dual space rendezvous, when the docked spacecraft and Agena were manoeuvred to the still orbiting Gemini 8 Agena target, so that Astronaut Collins could spacewalk across to it and retrieve a device fitted to record the effect of micro-meteoroid bombardment

ATLAS—US LAUNCHER

Lift-Off Weight:	260,000 lb (117,934 kg)
Lift-Off Thrust:	367,000 lb (166,460 kg)
Height (with Mercury):	93 ft (28·35 m)
Height (as Missile):	79 ft (24·08 m)
Diameter:	10 ft (3·05 m)
Speed at Burn-Out:	17,500 mph (28,166 kph)

History It was on top of an Atlas D that John Glenn, on February 20, 1962, became the first American in orbit. There were ten Atlas D flights in Project Mercury, the pioneering US manned programme; five unmanned test flights were followed by MA 5 with a chimpanzee, and four manned flights starting with Glenn. Technical problems with the rocket delayed Glenn's flight; many believed it would never be reliable enough for manned flight. In fact all four manned flights were successful, and the rocket has now been used in more than twenty-five space programmes. By the end of 1971 it had been fired nearly 400 times. It has sent Ranger spacecraft to the moon, and Mariners to Mars and Venus in combination with Agena as an upper stage. Originally developed as America's first US ICBM (Intercontinental Ballistic Missile), it became operational in September 1959; a total of 159, with a range of over 5000 miles (8050 km), were at one time deployed at sites across the US. Later versions had a range of 8000 miles (12,875 km).

General Description Atlas-D is 65 ft (20 m) from base to Mercury adapter section, and 10 ft (3·05 m) diameter at tank section. Construction is of such thin-gauge metal that it must be pressurized during ground transport, and as its propellants are consumed during flight, to maintain structural rigidity. All five engines, which burn highly refined RP–1 kerosene and liquid oxygen, are fired for take-off: the central sustainer engine (60,000 lb; 27,215 kg thrust), the 2 outer booster engines (150,000 lb; 68,040 kg thrust each), and the 2 small vernier engines used for mid-course corrections during powered flight. System components, including command receivers, telemetry, guidance, antennae etc., are in pods on the side of the fuel tank located immediately above the main engines. Launch-vehicle guidance is provided

145

left: Atlas launch—the vehicle used to place John Glenn in orbit

right: Redstone sending Alan Shepard on the first US suborbital mission

by a combination of on-board and radio ground guidance equipment. At T + 130 seconds, ground command shuts off the booster engines, and ground guidance controls the sustainer engine; the boosters are jettisoned at shut-off, taking with them the large flared 'skirt' around the tail, which provides stability during the initial flight stages. At about T + 300 seconds, when insertion parameters are attained, ground control shuts off sustainer and vernier engines. Spacecraft separation is at an altitude of about 40 miles (64·5 km), range of about 45 miles (72·5 km) from the launchpad, and velocity of 17,543 mph (28,232 kph).

REDSTONE—US LAUNCHER

Lift-off Weight:	66,000 lb (29,935 kg)
Lift-off Thrust:	78,000 lb (35,375 kg)
Height (with Mercury):	83 ft (25·30 m)
Height (as Missile):	69 ft 4 in. (21·13 m)
Diameter:	5 ft 10 in. (1·77 m)
Speed at Burn-out:	4400 mph (7080 kph)
Range:	200 miles (320 km)

History Redstone was used to launch the two suborbital flights which inaugurated America's manned space programme. It was first developed by Dr Wernher von Braun for the US Army from his original V2 missile; with a 600-mile (965-km) range, it was deployed in Europe in 1958, and became known as 'Old Reliable'. Some 800 engineering changes were made to transform it into a booster for the first experimental manned flights. In addition to major engine improvements, 20 seconds were added to the burn-time by lengthening the tank section by 6 ft (1·83 m). This, with the spacecraft and escape tower, added 5000 lb (2265 kg) to the original 61,000 lb (27,670 kg) missile weight. Telemetry provided sixty-five measurements covering attitude, vibration, acceleration, temperature, thrust level etc. Redstone also became established as a space launcher in 1958 when used as the first stage of the Jupiter C rocket which put Explorer 1, America's first satellite, into earth orbit.

Six Mercury-Redstone flights were made, starting in November 1960. The first four were test-firings; then came the historic 15-minute 'lob' which made Alan Shepard America's first man in space, on May 5, 1961, followed by Virgil Grissom on July 21, 1961. Details of these flights appear under **Mercury** (q.v.).

SATURN 1B—US LAUNCHER

Lift-Off Weight
  (with CSM):   1,295,600 lb (587,675 kg)
Overall Height: 224 ft (68·28 m)
Lift-Off Thrust: 1·6 million lb (0·73 million kg)
Payload: 40,000 lb (18,145 kg)
Diameter—1st Stage: 21 ft 5 in. (6·53 m)
             2nd Stage: 21 ft 8 in. (6·60 m)

History The Wernher von Braun organization, then working with the Army Ballistic Missile Agency, first proposed the need for a launcher of 1·5 million lb (680,400 kg) thrust, able to place between 10 and 20 tons (10·15 and 20·3 tonnes) in earth orbit, or send 3 to 10 tons (3·05 to 10·15 tonnes) on escape missions, in April 1957. By 1959 the project had been named Saturn. This was because Saturn was the next planet after Jupiter in the solar system, and the Saturn rocket

was the next von Braun project following completion of Jupiter missile development. Saturn 1, with engines and tanks in clusters in order to make use of equipment already developed, was a two-stage vehicle. The first stage had eight H-1 engines, burning RP-1 kerosene and liquid-oxygen, each generating 188,000 lb (85,250 kg) thrust. The second stage (designated S4), had six liquid-oxygen, liquid-hydrogen RL-10 A-3 engines, each generating 15,000 lb (6800 kg) thrust. Ten Saturn 1s were fired between October 1961 and July 1965, with an unprecedented record of 100% success. The fifth placed a 37,700 lb (17,100 kg) payload into earth orbit, and the sixth and seventh each placed unmanned 'boilerplate' models of Apollo spacecraft in earth orbit. The ninth orbited a Pegasus meteoroid technology satellite. Meanwhile, it had been decided that elements of Saturn 1 and of the planned, much larger Saturn 5, should be combined to form a new mid-range vehicle, Saturn 1B. This would have a 50% greater capability than Saturn 1, and enable complete Apollo spacecraft to be tested in manned earth orbital flights one year earlier than would be possible with Saturn 5.

149

General Description Saturn 1B's first stage retains the same size and diameter, and is 80 ft 4 in. (24·49 m) high, with 21 ft 5 in. (6·53 m) diameter; but its weight was reduced, and payload correspondingly increased by 20,000 lb (9070 kg), by a new fin design, and by resizing and redesigning tail section and spider beam etc. The eight H-1 engines have been uprated to 200,000 lb (90,700 kg). In 150 seconds from lift-off, they burn 42,000 gals (158,980 litres) of RP-1 fuel, and 67,000 gals (253,615 litres) of liquid oxygen, to reach an altitude of 42 miles (67·6 km) at burn-out.

The second stage, an enlarged S4, designated S4B, 58 ft 5 in. (17·81 m) high, and 21 ft 8 in. (6·60 m) in diameter, has a single liquid-hydrogen, liquid-oxygen J2 engine, giving 200,000 lb (90,700 kg) thrust. It burns 64,000 gals (242,250 litres) of liquid-hydrogen, and 20,000 gals (75,700 litres) of liquid-oxygen, in 450 seconds of operation, to achieve orbital speed and altitude. The Instrument Unit (IU), only 3 ft (0·91 m) high, with 21 ft 8 in. (6·60 m) diameter, is known as 'the wedding ring', since it joins launcher and spacecraft. Unpressurized, it houses the instrumentation concerned with vehicle performance from lift-off to insertion of payload into orbit; it contains tracking command, measuring and telemetry systems, an electrical power supply and thermal conditioning system, and in the Skylab mission is designed for a $7\frac{1}{2}$-hour life after lift-off. Throughout the launch the IU systems measure the vehicle's rate of acceleration and attitude, calculate what corrections are needed to keep it on course, and issue commands to the engines, shortening or lengthening their burn-time, so that the vehicle achieves the exact height and speed needed for the mission.

Five Saturn 1B firings have maintained Saturn

1's record of success. The first launch, which was also the first flight test of a powered (unmanned) Apollo spacecraft, was on February 26, 1966; after three more test flights, the first manned flight of the series, Apollo 7, was successfully launched on October 11, 1968. Three Saturn 1B's will be used to send the three rotating crews of astronauts to the Skylab space station in 1973. Of the twelve originally built, four will then remain in long-term storage for use in future missions. The Saturn 1 programme cost £319,580,000 ($767 million); Saturn 1B an additional £469,587,000 ($1127 million).

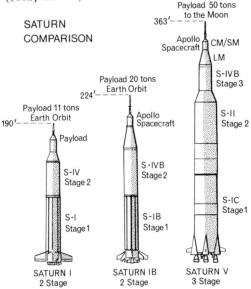

SATURN COMPARISON

Payload 50 tons to the Moon

363'

Apollo Spacecraft — CM/SM
LM
S-IVB Stage 3
S-II Stage 2
S-IC Stage 1

SATURN V
3 Stage

Payload 20 tons Earth Orbit

224'

Apollo Spacecraft
S-IVB Stage 2
S-IB Stage 1

SATURN 1B
2 Stage

Payload 11 tons Earth Orbit

190'

Payload
S-IV Stage 2
S-I Stage 1

SATURN 1
2 Stage

SATURN 5—US LAUNCHER

Lift-Off Weight (Apollo 11):	6,477,875 lb (2,938,312 kg)
First-Stage Thrust (Apollo 11):	7,552,000 lb (3,425,500 kg)
Height:	281 ft (86 m)
(with Apollo)	363 ft (111 m)
(with Skylab)	357 ft (109 m)
Diameter: (1st and 2nd stages):	33 ft (10·06 m)
(3rd stage):	21 ft 8 in. (6·60 m)

History Saturn 5, designed and developed at NASA's Marshall Space Flight Centre under Dr Wernher von Braun, for the Apollo Project (Plate 4), will also be used to launch the Skylab orbiting workshop. Fifteen were built, and when Apollo 17 concludes the moonlanding project in 1972, three will remain for future projects such as Skylab.

Saturn 5 has had an impressive record of success since development began in January 1962. Its first flight was on November 9, 1967, the unmanned Apollo 4 mission, which successfully tested both the rocket and the spacecraft. It was first used for a manned flight on the Apollo 8 mission, on December 21, 1968, when Frank Borman commanded man's first historic flight around the moon. Its flexibility was demonstrated when it survived a lightning strike during the Apollo 12 launch; and on Apollo 13, when the centre engine of stage two shut down 2 minutes early. The other four engines automatically burned to depletion to make up the lost thrust, and the third stage automatically burned for an extra 10 seconds to complete the task. There was still ample fuel left to complete

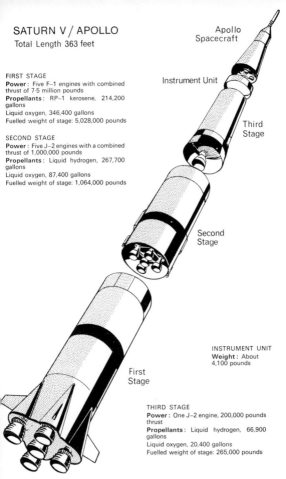

SATURN V / APOLLO
Total Length 363 feet

Apollo Spacecraft

Instrument Unit

Third Stage

Second Stage

First Stage

FIRST STAGE
Power : Five F–1 engines with combined thrust of 7·5 million pounds
Propellants : RP–1 kerosene, 214,200 gallons
Liquid oxygen, 346,400 gallons
Fuelled weight of stage: 5,028,000 pounds

SECOND STAGE
Power : Five J–2 engines with a combined thrust of 1,000,000 pounds
Propellants : Liquid hydrogen, 267,700 gallons
Liquid oxygen, 87,400 gallons
Fuelled weight of stage: 1,064,000 pounds

INSTRUMENT UNIT
Weight : About 4,100 pounds

THIRD STAGE
Power : One J–2 engine, 200,000 pounds thrust
Propellants : Liquid hydrogen, 66,900 gallons
Liquid oxygen, 20,400 gallons
Fuelled weight of stage: 265,000 pounds

the translunar injection. Structural weight reductions, with improved engine performance and operational techniques, enabled the Saturn 5s used in the final Apollo missions to place payloads of 150 tons (152·4 tonnes) into earth orbit and to send 53 tons (53·9 tonnes) to the moon.

General Description *First Stage* (S1C) Built by the Boeing Co. at NASA's Michoud Assembly Facility, New Orleans. Its five F-1 engines consume kerosene and liquid oxygen at 29,364 lb (13,319 kg) per second, and must boost the vehicle to approximately 5300 mph (8530 kph) and a height of 38 miles (61 km) in 2½ minutes. Major components are the forward skirt, oxidizer tank, intertank structure, fuel tank and thrust structure. One engine is rigidly mounted on the stage's centreline; the other four, mounted on a ring at 90° angles around the centre engine, are gimballed to control the vehicle's attitude during flight.

Second Stage (S2) Built by the Space Division of North American Rockwell corporation at Seal Beach, California. Its five J-2 engines ignite as the first stage separates and falls away, and develop a total thrust of 1,160,000 lb (526,165 kg), burning a mixture of liquid hydrogen and liquid oxygen. They must raise the vehicle's speed to approximately 15,300 mph (24,625 kph), and a height of 114 miles (183·5 km) in 6 minutes. Major components are the forward skirt, liquid hydrogen and liquid oxygen tanks (separated by an insulated common bulkhead), a thrust structure, and an interstage section connecting it with the first stage. As on the first stage, the centre engine is rigid, and the outer four can be gimballed.

Third Stage (S4B)   Built by the McDonnell Douglas Astronautics Co. at Huntingdon Beach, California.   The function of this stage is quite different.   Its single, J-2, gimballed engine, powered by liquid hydrogen and liquid oxygen, has a maximum thrust of 230,000 lb (104,325 kg) and can be shut off and re-started.   Its first job is to take over when stage two falls away, and burn for about 2½ minutes (the exact time is controlled by computer) to increase the speed to the necessary orbital rate of 17,400 mph (28,000 kph). Major components are the aft interstage and skirt, thrust structure, two propellant tanks with common bulkhead and forward skirt.   On a moon-flight, the S4B shuts down after placing the vehicle into a parking orbit, and remains attached to the spacecraft.   Usually on the second orbit is it fired a second time for approximately 5 minutes to accelerate the vehicle to over 24,400 mph (39,270 kph), thus injecting it into a translunar orbit. Shortly after (about 3 hours after lift-off), the Apollo Command Module separates from the nose of the S4B, turns around, docks and withdraws the Lunar Module, protectively housed for take-off just below the Command Module, and inside the top of the S4B.   This done, the S4B is either sent off into solar orbit to ensure that its path is well clear of Apollo, or it is sent on to impact at a selected point on the moon's surface, for seismometer and other scientific tests.

Instrument Stage   As described under **Saturn 1B** (q.v.)

Total Saturn 5 costs were £2,519,166,000 ($6,046 million); including Saturns 1 and 1B, expenditure on Saturn programmes totals £3,308,300,000 ($7,940 million).

SOYUZ—USSR LAUNCHER

Lift-Off Thrust: 1,124,000 lb (509,840 kg)
Upper Stage: 309,000 lb (140,160 kg)
Height: 140 ft (42·65 m)

General Description Soyuz figures are estimates. This launcher appears to be a development of the Vostok and Voskhod rockets, consisting mainly of the insertion of 36 ft (11·8 m) of additional upper staging, with strengthening of the inter-stage truss to safeguard against bending. An emergency escape tower is mounted above the spacecraft, with three separate tiers of rocket motors; in the event of a launchpad abort, these systems boost the spacecraft clear of the launchpad, bend the trajectory and control the recovery. The Soyuz spacecraft usually weigh just over 13,225 lb (6000 kg).

These launchers, with an orbital 'escape' stage combined with the upper stage, are also used for lunar, Venus and Mars probes, and for placing Molniya-type communications satellites into earth orbit.

Perhaps the most remarkable thing about Soviet manned spaceflight is the small amount of rocket development which appears to have taken place between the first Vostok mission and the Soyuz series, when compared with the 367,000 lb (166,460 kg) thrust of the Atlas rocket used for the first US orbital flight, and the 7,550,000 lb (3,400,000 kg) thrust of the Saturn 5 used for Apollo.

left: Soyuz 11 on its record **twenty-three-day** but **ultimately** ill-fated mission

right: Vostok—the launcher used for Russia's first six spaceflights

TITAN 2—US LAUNCHER

Lift-Off Thrust:	430,000 lb (195,000 kg)
Second Stage Thrust:	100,000 lb (45,360 kg)
Lift-Off Weight:	327,000 lb (148,325 kg)
Height with Spacecraft:	109 ft (33·22 m)
Height as Missile:	103 ft (31·40 m)
1st Stage Diameter:	10 ft (3·05 m)
Length of 1st Stage:	70 ft (21·34 m)

History   Titan 2 was successfully used to launch all ten of the two-man Gemini spacecraft which followed the Mercury-Atlas flights.   Fifteen Titan 2s, adapted as space launchers from the original Titan series of ICBMs, were ordered by NASA, and on April 8, 1964 an unmanned Gemini was put into orbit at the first attempt. From the start Titan proved to be one of the most dependable US missiles; Titan I had only four failures in forty-seven launches: Titan 2, first launched in 1962, had a range of over 6000 miles (9655 km), and its storable propellants gave it a capability of being launched from underground silos in less than a minute.   Titan's reliability enabled Project Gemini—completed in only 20 months—to mark the start of sophisticated manned spaceflight; this was demonstrated by Gemini 7 and 6 being launched from the same pad within 11 days; the craft were regularly placed in such precise orbits that it was possible in less than 2 years to master the technique of rendezvous and docking.

General Description   A two-stage vehicle, with a rigid structure of high-strength aluminium, Titan has the advantage over Atlas of not needing to be pressurized to maintain its rigidity on the launchpad; the fuel-tank walls double as the outer skin of the vehicle. Both stages are liquid-fuelled; they burn unsymmetrical dimethyl-hydrazine (UDMH), with nitrogen tetroxide as oxidizer. The fuels are storable and hypergolic (that is, they ignite on contact, so that an ignition system is unnecessary). 'Fire-in-the-hole staging' is employed —that is, second-stage engine ignites before separation from first stage is complete at Lift-off + 2 minutes 30 seconds. Second-stage engine shuts down at $5\frac{1}{2}$ minutes, at 100 miles (161 km) altitude, and 531 miles (854·5 km) downrange; spacecraft separation follows at 5 minutes 50 seconds. Most important of the modifications, first installed on the man-rated Titans, is the Malfunction Detection Systems (MDS). This continually monitors performance of the sub-systems, and signals reports to the astronauts, thus enabling them to decide whether the mission should be continued or aborted.

159

VOSKHOD—USSR LAUNCHER

Estimated $\begin{cases}$ Total Thrust: & 1,433,000 lb (650,000 kg) \\ Height: & 140 ft (42·65 m) $\end{cases}$

General Description A developed version of the Vostok launcher. The first stage is probably almost identical with Vostok's although there is some evidence that the propellant tanks have been enlarged.

The upper stage has been lengthened to about 20 ft (6 m) compared with 6·5 ft (1·98 m) on Vostok. The Russians say that it contains seven engines instead of six. The second stage, therefore, is believed to have a twin-chamber rocket engine, developing about 308,650 lb (140,000 kg) thrust.

Since ejection seats were omitted from the Voskhod spacecraft, an emergency escape tower was added to the rocket for the launch of Voskhods 1 and 2.

VOSTOK—USSR LAUNCHER

Lift-Off Thrust:	1,124,000 lb (509,840 kg)
Upper Stage Thrust:	199,000 lb (90,265 kg)
Total Height:	124 ft 8½ in. (38 m)
Upper Stage and Fairing:	32 ft 10 in. (10 m)
Strap-on Boosters:	62 ft 3½ in. (19 m)
Overall Diameter:	34 ft (10·36 m)
Diameter of Boosters:	9 ft 10 in. (3 m)

General Description All figures are necessarily estimates. The Vostok rocket, not shown publicly until 6 years after it had been used to place Yuri Gagarin in orbit, is a cluster, or pyramid design, and consists of six units (Plate 8). The central unit

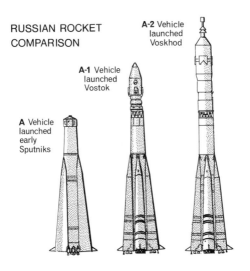

RUSSIAN ROCKET
COMPARISON

A-2 Vehicle
launched
Voskhod

A-1 Vehicle
launched
Vostok

A Vehicle
launched
early
Sputniks

with four primary nozzles and four verniers (small swivelling nozzles), is supported by four clip-on boosters, each with four primary and two vernier nozzles. Thus, no fewer than thirty-two rocket chambers must be fired simultaneously for lift-off. The upper stage (referred to as the third stage by Soviet scientists) has one primary and four vernier, or steering nozzles. The central unit is a slim cylinder which flares out into a larger diameter at the top, thus enabling the clip-on units to fit snugly against it; they are fastened by two belts of explosive bolts at the top and bottom, and are jettisoned when burnt out. The central unit continues firing at full thrust (224,870 lb; 102,000 kg), and, when through the densest layers of atmosphere, the protective nose cone is jettisoned. The upper stage ignites at burn-out of the central unit, which then separates. This cuts out when the required orbital velocity is achieved, and separates from the spacecraft. All the primary nozzles are fixed, control during launch being exercised by the vernier nozzles, supplemented by four small aerodynamic control surfaces (officially described as 'rudders'), at the base of each clip-on booster. Each rocket unit is stated to use an individual engine firing LOX (liquid oxygen) and hydrocarbons. Vostok can place a payload of about 10,400 lb (4715 kg) in earth orbit, not counting the upper stage, the burn-out weight of which is about 3200 lb (1450 kg). Vostok is assembled in a horizontal position attached to a 'strongback', then taken by rail to the Tyuratum launchpad, where it is raised to a vertical position over a flame deflector pit. It is steadied by supporting arms which swing up from the pad and remain there until lift-off, when they swing back again.

The secrecy surrounding the Vostok rocket was due to the fact that it was developed from an ICBM; it was designed by the late Sergei Korolev, and used for all six Vostok launchings.

SPACE CENTRES

BAIKONUR SPACE CENTRE

The Soviet equivalent of Cape Kennedy, it is in fact near Tyuratum, in Kazakhstan, east of the Aral Sea, about 230 miles (370 km) south-west of Baikonur. The exact position has always been the subject of as much secrecy as its equipment and operation. Though one or two Westerners have been admitted for satellite launches, at the time of writing no outsider has ever been present for a manned launch.

According to the *Soviet Encyclopaedia of Spaceflight*, the Baikonur range extends for thousands of kilometers over USSR territory, and terminates in the Pacific Ocean. It has 'a number of launch complexes, maintenance areas and tracking stations', with a number of tracking stations also located along its ranges. Sputnik 1, the world's first satellite, Yuri Gagarin, the first man in orbit, and all Russia's subsequent manned flights, were launched from Baikonur.

All the evidence suggests that the Russians have always used a mobile system for their launches, assembling the rockets and spacecraft first and then using transporters to take them to the pad, somewhat similar to America's Launch Complex 39. The system of 'clustered' engines, abandoned by America after Saturn 1, has so far been retained however, and probably explains why Russia's largest rocket is more comparable to Saturn 1 than to Saturn 5 in size and thrust.

MANNED SPACECRAFT CENTRE, HOUSTON

Control of all manned spaceflights passes from Launch Control Centre (LCC) at Cape Kennedy to Manned Spacecraft Centre (MSC), 20 miles (32 km) south-east of Houston, 10 seconds after lift-off. Until 1961 it was undeveloped cattle country; in 3 years it became the world's most sophisticated scientific centre. Unlike an airport, it was built as a whole, with no continuing process of additions and alterations. Its focal point, Mission Control Centre, has been the control point for all NASA's manned flights since Gemini 4, in June 1965. Flight controllers at 171 consoles, can check, at a touch of a button, the progress of the mission against the flight plan; 133 TV cameras and 557 receivers are used.

MSC is also responsible for design, development and testing of manned spacecraft and their systems, as well as the selection and training of astronauts, who live nearby in the Clear Lake and Key Lago areas.

KENNEDY SPACE CENTRE

Cape Canaveral (later renamed Cape Kennedy), a strip of sandy jungle on the Florida coastline, was originally set aside as a development area for ICBMs (Intercontinental Ballistic Missiles) in 1947. Since then it has been developed into the John F. Kennedy Space Centre, and all US manned spaceflights have been launched from it. About 40 Launch Complexes have been built, mostly by the US Air Force for their own military missile development and for unmanned satellite launch-

Vehicle Assembly Building at Kennedy Space Centre—in foreground spacecraft being transported to launch site

ings carried out by them for NASA. The Merritt Island 'Moonport', Launch Complex 39, however, though part of the Kennedy Centre, was developed and is operated exclusively by NASA for the Apollo moonlandings and subsequent missions such as Skylab.

Bus tours are operated by NASA for visitors at a rate of about £1 ($2.50) for adults, and half price for those under 18, and on Sundays visitors are allowed to drive their own cars to the launch sites.

Launch Complexes involved in manned space-flight are as follows:

5–6 Site of first two US suborbital manned flights, Mercury-Atlas 3 and 4. Complex now a museum piece.

14 Used for Mercury-Atlas 6–9, the four

orbital flights, starting with John Glenn in February 1962. Also disused, but preserved for visitors.

19 All ten Gemini-Titan flights launched from here, between 1965–66. Site now inactive.

34 Site of first Saturn 1 launch. Altogether seven Saturn 1 and 1B launches took place here, including Apollo 7, in October 1968, the project's first manned flight. The scene, too, of the Apollo 1 disaster; Grissom, White and Chaffee died when fire broke out in their spacecraft during what should have been a final rehearsal for the Apollo 7 launch. Site now inactive.

37 Site of eight Saturn 1 and 1B launches, but no manned flights. Inactive.

39 Site of all manned Apollo flights except the first (Apollo 7), and of Skylab launches. Its construction reversed the fixed launch concept of earlier missions, with assembly, checkout and launch all being carried out on the pad. With the mobile concept employed on Complex 39, assembly is carried out in the huge Vehicle Assembly Building (VAB), which is 525 ft (160 m) tall, and so large that four buildings the size of the United Nations in New York would fit inside. The Launch Control Centre (LCC) nearby has four firing rooms, each with 470 sets of control and monitoring equipment, plus conference and display rooms etc. During launch sixty-two TV cameras at the pad supply 100 monitor screens. The Mobile Launcher, 446 ft (136 m) high and 10·6 million lb (4808 tonnes) in weight, serves as an assembly platform within the VAB and as a launch platform and umbilical tower at the launch site 3½ miles (5·6 km) away. The Mobile Launcher is moved along a special roadway by the

Crawler-Transporter at Kennedy Space Centre

Crawler-Transporter, a double-tracked vehicle the size of a football field, 131 ft (40 m) long, and 114 ft (35 m) wide; the assembled Apollo-Saturn 5 vehicle which it carries stands thirty-six storeys high; the total load weighs 17 million lb (7711 tonnes).

Access to both the Apollo spacecraft and Saturn vehicle at the pad is provided by the Mobile Service Structure (MSS). 410 ft (125 m) high, this has two lifts and provides five platforms—the top three for spacecraft access, and the lower two for Saturn 5. About 11 hours before launch, the MSS is moved back 7000 ft (2134 m) from Pad A. Pads A and B are almost identical, although Pad B was used only once, for Apollo 10, up to the end of the Apollo Project. They are 8716 ft (2657 m) apart and roughly octagonal in shape. Below them are flame trenches 42 ft (13 m) deep and 450 ft (137 m) long. For Emergency, a 200-ft (61-m) escape tube leads from the bottom of the lifts to a blast-resistant room 40 ft (12 m) below the pad; a cab on a slidewire also provides quick escape from the 320-ft (98-m) level on the MSS to a revetment 2500 ft (762 m) away.

SPACEMEN

US ASTRONAUTS

At the end of 1971 forty-four astronauts were on
the active list out of a total of seventy-three pilots
and scientists selected in seven groups since April
1959. Of these, thirty-three were pilot-astronauts
and twelve were scientist-astronauts. Of the
twenty-nine no longer on the active list, eight were
dead; three were lost in an Apollo fire during a
ground rehearsal, but not one in space. Up to and
including Apollo 15, thirty astronauts had flown
and a further eleven with no previous flight
experience had been nominated for missions up
to the end of Skylab in 1973.

Brief details of all seventy-three appear below:

Aldrin, Edwin E. (b. Jan. 20, 1930): Col. USAF;
pilot-astronaut, selected Oct. 1963. Second man
to step on moon, as pilot of Apollo 11's 'Eagle' on
July 20, 1969. Previously flew in Gemini 12 and
spent $5\frac{1}{2}$ hrs on EVA outside spacecraft. Left
NASA 1971. Ret. USAF 1972. M; 2 sons, 1 dtr.

Allen, Joseph P. (b. June 27, 1937): Scientist-
astronaut, selected Aug. 1967; nuclear physicist.
M; 1 son.

Armstrong, Neil A. (b. Aug. 5, 1930): Civilian
test pilot; pilot-astronaut, selected Sep. 1962.
First man to step on moon, July 20, 1969 as Cdr
Apollo 11. As Cdr Gemini 8 on Mar. 16, 1966,

left: Edwin Aldrin   *right:* Neil Armstrong

performed first space-docking with Agena target.
Prof. Engineering, U. of Cincinnati, Ohio, Oct.
1971. M; 2 sons.

Bassett, Charles A. (b. Dec. 30, 1931); Maj.
USAF; pilot-astronaut, selected Oct. 1963. Was
training as Gemini 9 pilot when he and Elliott See
were killed in T-38 crash at St. Louis in Feb. 1966.

Bean, Alan L. (b. Mar. 15, 1932): Capt. USN;
pilot-astronaut, selected Oct. 1963. LMP on
Apollo 12, Nov. 1969 was 4th man on moon.
Nominated as Cdr Skylab 2 mission in 1973.   M;
1 son, 1 dtr.

Bobko, Karol J. (b. Dec. 23, 1937): Maj. USAF;
pilot-astronaut, transferred to NASA when AF
Manned Orbiting Laboratory programme was
cancelled in Aug. 1969. M; 1 son, 1 dtr.

Borman, Frank (b. Mar. 14, 1928): Col. USAF
(Ret.); pilot-astronaut, selected Sep. 1962. Cdr
Apollo 8, on first circumlunar flight, Dec. 1968.

Cdr Gemini 7 in first RV with Gemini 6 on Dec. 4, 1965. Sen. Vice-Pres., Eastern Airlines, July 1, 1970.   M; 2 sons.

Brand, Vance D. (b. May 9, 1931): Civilian test pilot; pilot-astronaut, selected Apr. 1966.   Support crew member on Apollos 8 and 13.   M; 2 sons, 2 dtrs.

Bull, John S. (b. Sep. 25, 1934): Lt-Cdr USN (Ret.); pilot-astronaut, selected Apr. 1966; withdrew, following illness, 1968.

Carpenter, M. Scott (b. May 1, 1925): Cdr USN (Ret.); Mercury astronaut, selected Apr. 1959; made 2nd orbital flight in Mercury 7, May 24, 1962.   Resigned NASA 1967.   Pres. Sea Sciences Corp., Los Angeles.   M; 4 children.

Carr, Gerald P. (b. Aug. 22, 1932): Lt-Col. USMC; pilot-astronaut, selected Apr. 1966. Support crew member for Apollos 8 and 12. Nominated Cdr Skylab 3, planned for Oct. 1973. M; 2 sons, 4 dtrs.

Cernan, Eugene A. (b. Mar. 14, 1932): Capt. USN; pilot-astronaut, selected Oct. 1963.   As LMP on Apollo 10, descended to within 9 miles of lunar surface in final landing rehearsal on May 18, 1969.   Nominated Cdr Apollo 17, Dec. 1972. M; 1 dtr.

Chaffee, Roger B. (b. Feb. 15, 1935): Lt-Cdr USN; pilot-astronaut, selected Oct. 1963.   Selected for Apollo 7, but died in launchpad fire, Jan. 27, 1967.

Chapman, Philip K. (b. Mar. 5, 1935): Civilian;

scientist-astronaut, selected Aug. 1967. Australian. M; 2 children.

Collins, Michael (b. Oct. 31, 1930): Col. USAF Res.; pilot-astronaut, selected Oct. 1963. As CMP of Apollo 11, in July 1969, remained in lunar orbit during first moonlanding. Pilot on Gemini 10 on July 18, 1966. Dir. Air and Space Mus., Smithsonian Inst., Washington, Feb. 1971. M; 1 son, 2 dtrs.

Conrad, Charles (b. June 2, 1930): Capt. USN; test-pilot astronaut, selected Sep. 1962. As Cdr Apollo 12, 3rd man to land on moon. Pilot of Gemini 5 on 8-day flight in Aug. 1965; Cdr Gemini 11, Sep. 12, 1966. Head of Skylab Project, and nominated Cdr Skylab 1, scheduled for May 1, 1973. M; 4 sons.

Cooper, L. Gordon (b. Mar. 6, 1927): Col. USAF (Ret.); Mercury astronaut, selected Apr. 1959. Made Mercury 9 flight, last of series on May 15, 1963. Cdr Gemini 5, Aug. 1965. Back-up Cdr

First Skylab crew: Charles Conrad, Joseph Kerwin and Paul Weitz

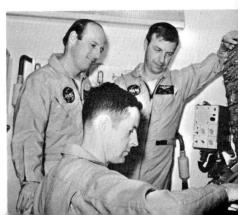

Apollo 10. Ret. July 1970; in business as Gordon Cooper Associates, Hialeah, Fla. M; 2 dtrs.

Crippen, Robert L. (b. Sep. 11, 1937): Lt-Cdr USN; pilot-astronaut, transferred to NASA when A.F. Manned Orbiting Laboratory programme was cancelled in Aug. 1969. M; 3 dtrs.

Cunningham, Walter (b. Mar. 16, 1932): Civilian; pilot-astronaut, selected Oct. 1963. LMP on Apollo 7 in Oct. 1968. Worked on Skylab. Resigned Aug. 1971. Vice-President of Operations, Century Development Co. Houston. M; 1 son, 1 dtr.

Duke, Charles M. (b. Oct. 3, 1935): Lt-Col. USAF; pilot-astronaut, selected Apr. 1966. LMP Apollo 16, Apr. 1972, and 10th man on moon. M; 2 sons.

Eisele, Donn F. (b. June 23, 1930): Col. USAF; pilot-astronaut, selected Oct. 1963. CMP on Apollo 7, Oct. 1968; back-up CMP Apollo 10; from 1970 Tech. Ass. Manned Spaceflight, Langley. M; 2 sons, 2 dtrs.

England, Anthony W. (b. May 15, 1942): Civilian; scientist-astronaut, selected 1967. M; 1 dtr.

Engle, Joe H. (b. Aug. 26, 1932): Lt-Col. USAF; pilot-astronaut, selected Apr. 1966. Support crew for Apollo 10; back-up LMP Apollo 14. M; 1 son, 1 dtr.

Evans, Ronald E. (b. Nov. 10, 1933): Cdr USN; pilot-astronaut, selected Apr. 1966. Support crew Apollo 7 and 11; back-up CMP, Apollo 14. Nominated CMP Apollo 17, Dec. 1972. M; 1 son, 1 dtr.

Freeman, Theodore C. (b. Feb. 18, 1930): pilot-astronaut, selected Oct. 1963. Killed in jet trainer crash, Oct. 1964.

Fullerton, Charles G. (b. 11 Oct. 1936): Maj. USAF; pilot-astronaut, transferred to NASA when AF MOL programme cancelled in Aug. 1969. M.

Garriott, Owen K. (b. Nov. 22, 1930): Civilian; scientist-astronaut selected June 1965. Nominated science-pilot Skylab 2, due in July 1973. M; 3 sons, 1 dtr.

Gibson, Edward G. (b. Nov. 8, 1936): Civilian; scientist-astronaut, selected June 1965. Support crew for Apollo 12; nominated science-pilot, Skylab 3 mission, due Oct. 1973. M; 1 son, 1 dtr.

Givens, Edward G. (b. Jan. 5, 1930): Maj. USAF. Died in car accident, June 1967.

Glenn, John H. (b. July 18, 1921): Col. USMC (Ret.); Mercury astronaut, selected April 1959. First US man in orbit, Mercury 6, on Feb. 20, 1962. Resigned NASA, 1964. On staff of Governor of Ohio. M; 1 son, 1 dtr.

Gordon, Richard F. (b. Oct. 5, 1929): Capt. USN (Ret.); pilot-astronaut, selected Oct. 1963. As CMP on Apollo 12, remained in lunar orbit during 2nd moonlanding, Nov. 1969. Pilot on Gemini 11, Sep. 12, 1966. Ret. for business career, Jan. 1972. M; 4 sons, 2 dtrs.

Graveline, Duane: Doctor; scientist-astronaut, selected June 1965; resigned for personal reasons, 1965.

Grissom, Virgil I. (b. Apr. 23, 1926): Lt-Col. USAF; Mercury astronaut, selected Apr. 1959; flew 2nd suborbital Mercury mission, July 21, 1961. Cdr Gemini 3, first US 2-man craft, Mar. 23, 1965. Died during Apollo launchpad rehearsal fire, Jan. 27, 1967.

Haise, Fred W. (b. Nov. 14, 1933): Civilian; pilot-astronaut, selected Apr. 1966. LMP on Apollo 13, Apr. 1970. Back-up Cdr Apollo 16, Apr. 1972. M; 2 sons, 1 dtr.

Hartsfield, Henry W. (b. Nov. 21, 1933): Lt-Col. USAF; pilot-astronaut, transferred to NASA when AF MOL programme cancelled, Aug. 1969. M; 2 dtrs.

Henize, Karl G. (b. Oct. 17, 1926): Astronomer; scientist-astronaut, selected Aug. 1967. Support crew, Apollo 15. M; 1 son, 2 dtrs.

Holmquest, Donald (b. Apr. 7, 1939): Doctor; scientist-astronaut, selected Aug. 1967. On leave as Asst. Prof. Radiology and Physiology Baylor Coll. of Medicine, Houston. M.

Irwin, James B. (b. Mar. 17, 1930): Col USAF; pilot-astronaut, selected Apr. 1966. As LMP on Apollo 15, July, 1971, was 8th man on moon, and drove first Lunar Rover. Back-up LMP for Apollo 17, Dec. 1972. M; 1 son, 3 dtrs.

Kerwin, Joseph P. (b. Feb. 19, 1932). Cdr USN;

Apollo 13 crew: James Lovell, John Swigert and Fred
Haise

scientist-astronaut, selected June, 1965. Nominated science-pilot for Skylab 1 mission on May
1, 1973. M; 3 dtrs.

Lenoir, William B. (b. Mar. 14, 1939). Civilian;
scientist-astronaut, selected Aug. 1967. Back-up
science-pilot for Skylabs 2 and 3, July and Oct.
1973. M; 1 son, 1 dtr.

Lind, Don L. (b. May 18, 1930): Cdr USN
(Res.); pilot-astronaut, selected Apr., 1966.
Back-up pilot, Skylab 2 and 3, July and Oct. 1973.
M; 2 sons, 3 dtrs.

Llewellyn, John A. (b. Apr. 22, 1933): Civilian;
scientist-astronaut, selected Aug. 1967. From
Cardiff, Wales, and only British-born astronaut
selected so far. Resigned for personal reasons,
Aug. 1968; Chemistry Prof. U. of Florida,
Tallahassee. M; 1 son, 2 dtrs.

Lousma, Jack R. (b. Feb. 29, 1936). Maj. USMC; pilot-astronaut, selected Apr. 1966. Support crew member, Apollo 9 and 13. Nominated pilot, Skylab 2, due July 30, 1973. M; 2 sons, 1 dtr.

Lovell, James A. (b. Mar. 25, 1928): Capt. USN; pilot-astronaut, selected Sep. 1962. First man to have flown 4 missions, with total of 715 hrs 4 mins in space. Cdr Apollo 13, April 1970, aborted moonlanding mission; pilot on Gemini 7, Dec. 4, 1965; Cdr Gemini 12, on Nov. 11, 1966; CMP Apollo 8, first circumlunar flight, Dec. 21, 1968. Since May 1971, Dep. Dir. Science and Applications, MSC, Houston. M; 2 sons, 2 dtrs.

Mattingly, Thomas K. (b. Mar. 17, 1936). Lt-Cdr USN; pilot-astronaut, selected Apr. 1966. Assigned as CMP to Apollo 13, but replaced following contact with German measles. CMP Apollo 16, Apr. 1972. M; 1 son.

McCandless, Bruce (b. June 8, 1937): Lt-Cdr USN; pilot-astronaut, selected Apr. 1966. Support crew member, Apollo 14; nominated back-up, pilot, Skylab 1, May 1973. M; 1 son, 1 dtr.

McDivitt, James (b. June 10, 1929): Brig-Gen. USAF; pilot-astronaut, selected Sep. 1962. As Cdr Apollo 9, achieved first LM/CSM docking. Cdr Gemini 4, June 3, 1965, when White made first US spacewalk. Since Sep. 1969, Manager Apollo Spacecraft Programme. M; 2 sons, 2 dtrs.

Michel, F. Curtis (b. June 5, 1934): Civilian; scientist-astronaut, selected June, 1965, resigned Aug. 1969, to return to scientific research, Rice U. Houston. M; 1 son, 1 dtr.

Mitchell, Edgar D. (b. Sep. 17, 1930): Capt. USN; pilot-astronaut, selected Apr. 1966. As LMP on Apollo 14, Feb. 1971, was 6th man on moon. Back-up LMP, Apollo 16. M; 2 dtrs.

Musgrave, Story (b. Aug. 19, 1935): Doctor; scientist-astronaut, selected Aug. 1967. Nominated back-up science-pilot, Skylab 1. M; 3 sons, 2 dtrs.

O'Leary, Brian T. (b. Jan. 27, 1940): Civilian; selected Aug. 1967. Resigned Apr. 1968. At U. of California, Berkeley. M.

Overmyer, Robert F. (b. July 14, 1936): Maj. USMC; pilot-astronaut transferred to NASA, when MOL programme was cancelled Aug. 1969. M; 2 dtrs.

Parker, Robert A. (b. Dec. 14, 1936): Civilian;

Apollo 9 crew: Russell Schweickart, James McDivitt and David Scott

scientist-astronaut, selected Aug. 1967. Support crew Apollo 15. M; 1 son, 1 dtr.

Peterson, Donald H. (b. Oct. 22, 1933): Maj. USAF; pilot astronaut transferred to NASA, when MOL programme was cancelled Aug. 1969. M; 1 son, 2 dtrs.

Pogue, William R. (b. Jan. 23, 1930): Lt-Col. USAF; pilot-astronaut, selected Apr. 1966. Support crew member Apollo 7, 11, 14. Nominated pilot, Skylab 3, due Oct. 28, 1973. M; 2 sons, 1 dtr.

Roosa, Stuart A. (b. Aug. 16, 1933): Lt-Col. USAF; pilot-astronaut, selected Apr. 1966. CMP on Apollo 14, Jan. 1971. M; 3 sons, 1 dtr.

Schirra, Walter M. (b. Mar. 12, 1923): Capt. USN (Ret.); Mercury astronaut, selected Apr. 1959. Pilot Mercury 8, Oct. 1962. Cdr Apollo 7, Oct. 1968. Cdr Gemini 6, Dec. 15, 1965 and achieved first RV with Gemini 7. Only man to fly Mercury, Gemini and Apollo missions. Ret. July 1969; Chairman Environmental Control Co., Denver. M; 1 son, 1 dtr.

Schmitt, Harrison H. (b. July 3, 1935): Civilian; scientist-astronaut, selected June, 1965. Back-up LMP Apollo 15. First scientist nominated for moon-landing as LMP Apollo 17, Dec. 6, 1972. Unmarried.

Schweickart, Russell L. (b. Oct. 25, 1935): Civilian; pilot-astronaut, selected Oct. 1963. As LMP, Apollo 9, in Mar. 1969, carried out first Apollo EVA. Nominated back-up Cdr Skylab 1, in 1973. M; 2 sons, 3 dtrs.

Scott, David R. (b. June 6, 1932): Col. USAF; pilot-astronaut, selected Oct. 1963. As Cdr Apollo 15, July 1971, was 7th man on moon. Pilot Gemini 8 on Mar. 16, 1966 for world's first docking. CMP Apollo 9, Mar. 1969 for first docking of CSM and LM. Nominated back-up Cdr Apollo 17, Dec. 1972. M; 1 son, 1 dtr.

See, Elliot M. (b. July 23, 1927): Civilian; pilot-astronaut, selected Sep. 1962. Was training as Cdr Gemini 9, when he and Charles Bassett were killed in T-38 crash at St Louis in Feb. 1966.

Shepard, Alan B. (b. Nov. 15, 1923): Adm. USN; Mercury astronaut, selected Apr. 1959. First American in space, on Mercury 3 suborbital flight, May 5, 1961. Lost flight status owing to ear trouble, and became Chief of Astronaut Office, 1963. Restored flight status, 1969; as Cdr Apollo 14, Feb. 1971, was 5th man on moon, and only Mercury astronaut to get there. Resumed duties as Chief of Astronaut Office. M; 2 dtrs.

left: Alan Shepard   *right:* Edward White

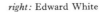

Slayton, Donald K. (b. Mar. 1, 1924): Civilian; Mercury astronaut, selected Apr. 1959. Lost flight status owing to heart condition. Co-ordinator, Astronaut Activities, Sep. 1962. From Nov. 1963, Director of Flight Crew Operations. Restored to flight status 1972. M; 1 son.

Stafford, Thomas (b. Sep. 17, 1930): Col. USAF; test-pilot astronaut selected Sep. 1962. Cdr Apollo 10 which in May 1969 flew within 9 miles of moon in final landing rehearsal. Pilot Gemini 6 Dec. 15, 1965 on world's first RV with Gemini 7. Cdr Gemini 9 on June 3, 1966. Dep. Dir. Flight Crew Operations. M; 2 dtrs.

Swigert, John L. (b. Aug. 30, 1931): Civilian; pilot-astronaut, selected Apr. 1966. CMP on aborted Apollo 13 moonlanding mission, Apr. 1970. Unmarried.

Thornton, William P. (b. Apr. 14, 1929): Doctor; scientist-astronaut, selected Aug. 1967. M; 2 sons.

Truly, Richard H. (b. Nov. 12, 1937): Lt-Cdr USN; pilot-astronaut, transferred to NASA when MOL programme was cancelled, Aug. 1969. M; 2 sons, 1 dtr.

Weitz, Paul L. (b. July 25, 1932): Cdr USN; pilot-astronaut, selected Apr. 1966. Support crew member, Apollo 12. Pilot, Skylab 1, due May 1, 1973. M; 1 son, 1 dtr.

Williams, Clifton C. (b. Sep. 26, 1932): Maj. USMC; pilot-astronaut, selected Oct. 1963. Killed in T-38 jet crash, Oct. 1967.

White, Edward H. (b. Nov. 14, 1930): Lt-Col. USAF; pilot-astronaut, selected Sep. 1962. As pilot Gemini 4 on June 3, 1965, made first US spacewalk. Died during Apollo launchpad rehearsal fire, Jan. 27, 1967.

Worden, Alfred M. (b. Feb. 7, 1932): Lt-Col. USAF; pilot-astronaut, selected Apr. 1966. CMP Apollo 15, 4th successful moonlanding, July 1971. Back-up CMP Apollo 17, Dec. 1972. Div.; 2 dtrs.

Young, John W. (b. Sep. 24, 1930): Cpt. USN; test-pilot-astronaut, selected Sep. 1962. CMP Apollo 10, final moonlanding rehearsal. Pilot Gemini 3, first US 2-man flight, Mar. 23, 1965; Cdr Gemini 10, July 18, 1966. As Cdr Apollo 16, Apr. 1972, became 9th man on moon, and 2nd to make 4 flights. M; 2 sons.

SOVIET COSMONAUTS

Russia does not announce names of her cosmonauts until they make a flight. By the end of 1971 twenty-five had flown in Russia's eighteen manned missions, including one woman. Of those, six were dead, four having been killed in re-entry accidents; of the other two Yuri Gagarin, the first man in space, was killed in a military aircraft accident, and the other died from natural causes. Full details of individuals are not available in every case.

Belyayev, Pavel I. (b. 1932): Cdr Voskhod 2, from which Leonov made world's first spacewalk, Mar. 18, 1965. Cosmonaut No. 10, he was also a poet and painter. Died from internal trouble, Jan. 11, 1970, aged 37.

Beregovoi, Giorgi T. (b. 1921): Maj-Gen; Cdr Soyuz 3, Oct. 1968, in RV flight with Soyuz 2. At 47 was oldest man to make first flight. Cosmonaut No. 12, joined cosmonaut team 1964.

Bykovsky, Valeri, F. (b. 1934): Pilot Vostok 5, June 1963, 2nd group flight, and was joined by Valentina Tereshkova, first woman in space, in Vostok 6. Cosmonaut No. 5.

Dobrovolsky, Georgi (b. June 1, 1928): Set new endurance record with 24-day flight in Soyuz 11 (23 days docked with Salyut) in June 1971. Killed by failure of hatch seal during re-entry. Cosmonaut No. 24, joined cosmonauts' team 1961. M; 2 dtrs.

Feoktistov, Konstantin P. (b. 1926): Scientist-cosmonaut on Voskhod 1, first 3-man craft, Oct. 1964. Cosmonaut No. 8, accepted for cosmonauts' group despite wartime injuries (shot and left for dead when acting as scout in German-occupied territory).

Filipchenko, Anatoli (b. 1928): Lt-Col; Cdr Soyuz 7, in record 3-craft groupflight with Soyuz 6 and 8, Oct. 1969. Cosmonaut No. 19. M; 2 sons.

Gagarin, Yuri A. (b. Mar. 9, 1934): Col. World's first man in space, in Vostok 1, Apr. 12, 1961. Subsequently became Cdr of Soviet Cosmonauts' Detachment, and made world-wide public appearances. Killed with Col. Seryogin during test flight of military jet, Mar. 27, 1968.

Ashes buried in Kremlin Wall; crater on lunar farside named after him.   M; 2 dtrs.

Gorbatko, Viktor (b. Dec. 3, 1934): Lt-Col; research engineer, Soyuz 7, group-flight with Soyuz 6 and 8, Oct. 1969. Cosmonaut No 21. M; 2 dtrs.

Khrunov, Yevgeny V. (b. 1933): Lt-Col; research engineer, Soyuz 5, Jan. 15, 1969; first docking of 2-manned craft, spacewalked with Yeliseyev to return in Soyuz 4. Cosmonaut No. 15, joined cosmonauts' team 1960. M; 1 son.

Komarov, Vladimir M. (b. Mar. 16, 1927): Col; Cdr Voskhod 1, first 3-man craft, on Oct. 12, 1964. Flew twice despite heart condition similar to Slayton's. First man to be killed in space, when re-entry parachute snarled at end of Soyuz 1, Apr. 23, 1967. Cosmonaut No. 7.

Andrian Nikolayev and Valentina Nikolayeva-Tereshkova with their daughter Lena

Kubasov, Valeri (b. 1935): Flt engineer, Soyuz 6, on Oct. 11, 1969 involving manoeuvres with Soyuz 7 and 8. Kubasov carried out 1st space-welding experiments. Cosmonaut No. 18, joined cosmonauts' team 1966. M; 1 dtr.

Leonov, Alexei A. (b. 1934): Col. First man to walk in space, on Voskhod 2, Mar. 18, 1965. Afterwards made vivid impressionistic paintings. Cosmonaut No. 11.

Nikolayev, Andrian G. (b. 1929): Col; pilot, Vostok 3, in 1st double flight with Vostok 4, Aug. 1962. Cdr Soyuz 9, which in June 1970 broke US endurance record with flight of nearly 18 days. Cosmonaut No. 3, joined team in 1960. Married V. Tereshkova, 1st space woman, in 1963; 1 dtr.

Patsayev, Viktor (b. June 19, 1933): Test engineer, Soyuz 11, on 24-day flight in Soyuz 11 and Salyut space station, June 1971. Killed following hatch-seal failure during re-entry. Cosmonaut No. 25, M: 1 son, 1 dtr.

Popovich, Pavel R. (b. 1930): Col; pilot, Vostok 4, in 1st double flight with Vostok 3, Aug. 1962. Cosmonaut No. 4, believed to be first appointment to cosmonauts' team.

Rukavishnikov, Nikolai (b. 1932): Test engineer, Soyuz 10, which docked briefly with Salyut, Apr. 1971. Cosmonaut No. 23, joined cosmonauts' team Jan. 1967. M; 1 son.

Sevastyanov, Vitali (b. July 8, 1935): Flt engineer, Soyuz 9, on 18-day endurance flight. Cosmonaut No. 22, joined team 1967. M; 1 dtr.

Soyuz 10 crew: Vladimir Shatalov, Alexei Yeliseyev and
Nikolai Rukavishnikov

Shatalov, Vladimir A. (b. 1927): Maj-Gen; Cdr
Soyuz 4, Jan. 1969, for first docking, with Soyuz
5, of 2-manned craft.   In Soyuz 8, Oct. 1969, was
Overall Cdr of Soyuz 6, 7 and 8 groupflight with
7 cosmonauts.   Cdr Soyuz 10, Apr. 1971, for
docking with Salyut.   Cosmonaut No. 13, joined
group 1963.   M; 1 son, 1 dtr.

Shonin, Georgi (b. 1935): Lt-Col; Cdr Soyuz 6,
on groupflight with Soyuz 7 and 8, Oct. 1969.
Cosmonaut No. 17.   M; 1 son, 1 dtr.

Tereshkova, Valentina V. (b. 1937): As pilot,
Vostok 6, on June 16, 1963, became first woman in
space in group flight with Vostok 5.   Cosmonaut
No. 6, she said that when Titov flew, she was 'not
even dreaming of becoming a cosmonaut', but
was in orbit 21 months after Titov.   Married
Nikolayev 1963; 1 dtr.

Titov, Herman S. (b. 1935): Col. As pilot Vostok

2, was second man in orbit, and 1st to spend one day there, on Aug. 6, 1961. Cosmonaut No. 2.

Volkov, Vladislav (b. Nov. 23, 1935): Flt engineer, Soyuz 7, on group flight with Soyuz 6 and 8, Oct. 1969. Flt engineer on 24-day flight in Soyuz 11 and Salyut space station, June 1971. Killed following hatch-seal failure during re-entry. Cosmonaut No. 20, joined group 1966. M; 1 son.

Volynov, Boris V. (b. 1934): Col; Cdr Soyuz 5, for first docking of 2-manned craft with Soyuz 4, Jan. 1969. Cosmonaut No. 14, joined group 1960. M; 1 son, 1 dtr.

Yegorov, Boris B. (b. 1937): First space doctor, on 3-man Voskhod 1, Oct. 1964. Cosmonaut No. 9.

Yeliseyev, Alexei S. (b. 1934): Flt engineer, Soyuz 5, for first docking of 2-manned craft with Soyuz 4, Jan. 1969. Spacewalked from Soyuz 5, and returned in Soyuz 4. Flt engineer on Soyuz 8, for group flight with Soyuz 6 and 7, Oct. 1969. Flt engineer on Soyuz 10, Apr. 1971. Cosmonaut No. 16, joined group 1966. M; 1 dtr.

INDEX